Electrochemistry

Electrochemistry

Wesley R. Browne

OXFORD
UNIVERSITY PRESS

OXFORD
UNIVERSITY PRESS

Great Clarendon Street, Oxford, OX2 6DP,
United Kingdom

Oxford University Press is a department of the University of Oxford.
It furthers the University's objective of excellence in research, scholarship,
and education by publishing worldwide. Oxford is a registered trade mark of
Oxford University Press in the UK and in certain other countries

Impression: 1

Published in the United States of America by Oxford University Press
198 Madison Avenue, New York, NY 10016, United States of America

British Library Cataloguing in Publication Data
Data available

Library of Congress Control Number: 2018951494

ISBN 978-0-19-879090-7

Printed in Great Britain by
Bell & Bain Ltd., Glasgow

Contents

Preface

Electrochemistry is one of the oldest and most interdisciplinary of the sciences; it is all around us and indeed within us, working unnoticed as you think about the text on this page. Yet it is often hidden to even those active in science and engineering under other research headings. Its modern application is seen in areas as diverse as energy, biology and biochemistry, healthcare, sensors, materials science, and, in addition to chemistry, the area draws heavily from physics and mathematics.

In writing this book, my aim is to provide a first introduction to electrochemistry to both undergraduates and those who have turned to this wonderful area of science at a later stage. Inevitably an understanding of the fundamentals of electrochemistry, and the 'tools of the trade' that are used, is essential in order to apply electrochemistry in other areas of science and technology. However, this is not an insurmountable challenge. Indeed, electrochemistry is a remarkably simple field building essentially on three relations:

$$\Delta G = -\nu FE = -RT \ln Q$$

$$\Delta G = \Delta H - T\Delta S$$

$$\Delta G = \Delta G^{\circ} + RT \ln Q$$

And it is from these equations that we build on throughout this book.

The book is comprised of six chapters, the first two of which deal with basic concepts in electronics and redox reactions, respectively. The aspects discussed include the concept of movement of charge and the relations that are encountered regularly in applied electrochemistry; of which, perhaps, of the most paramount importance, is the Nernst equation. Attention is given in Chapter 1 to electronic components, which although cursory, is nevertheless essential to understand concepts developed in later chapters such as impedance spectroscopy and the infamous '*iR* drop' or uncompensated resistance.

In Chapter 3, we explore the electrode-solution interface and how redox active species in solution affect electrode potentials before delving into the fundamental aspects of electrochemistry in Chapter 4, in which the basic equations are developed from first principles.

The 'tools of the trade' of the field of electroanalytical chemistry are introduced in Chapter 5 and especially the focus is on linear sweep and cyclic voltammetry, with a few excursions into application of electroanalytical chemistry to solve problems in molecular chemistry, in studying self-assembled monolayers, glucose sensing, and electrodimerizations.

In the final chapter, Chapter 6, the focus will shift to an important role played by electrochemistry in energy applications with batteries and fuel cells. The challenges faced in building and characterizing devices as well as their application will be discussed followed by an introduction to the world of impedance spectroscopy. Naturally, the coverage of all these areas cannot achieve

For didactical purposes, figures are simplified and, in many cases, are more cartoon than reality. These figures can be drawn exactly using spreadsheet or mathematical programs, which is a useful exercise. Furthermore, in some aspects, such as representing the stagnant layer as having a well defined thickness, generalizations that are not always valid are made so as to focus on the underlying concepts.

exhaustive depth; however, the primary goal is to demonstrate how important the basic principles are in getting a head start into more specialized fields.

The composition of the book and the order is based on my own experience in teaching electrochemistry at the undergraduate level and in more intense courses given at Masters and PhD level. The focus is on covering the fundamentals and providing the intellectual tools needed to tackle more advanced treatments, not least that of A. J. Bard and L. R. Faulkner—*Electrochemical methods: fundamentals and applications,*[1] as well as the books recommended in the further reading section.

Before beginning, there are five simple statements, paraphrased here, that were made by L. R. Faulkner[2] over 30 years ago and still hold true today.

i) Although we often focus on the solution containing our analyte of interest, an electrochemical system is much more than this. Several phases are in contact with each other, that is, a solid or liquid electrode, the electrolyte, membranes, salt bridges, and so on. Furthermore the bulk concentration of species in an electrolyte is not necessarily (or even usually) the same as that at the solid liquid interface (the Nernst diffusion layer). The redox reactions we are typically interested in only take place within a short distance (nanometres) of the electrode/solution interface and not in the bulk solution as would be the case for stoichiometric chemical oxidations.[3] To add to these aspects, diffusion and convection currents, stirring etc. will result in local inhomogeneities also. This aspects may seem to present an insurmountably complicated problem, but in fact we have a fairly solid understanding of these complex systems and the state of the art in electroanalytical techniques provides us with a myriad of approaches to figure out what is actually going on and where and when processes are actually happening.

ii) Many processes occur simultaneously. We will meet this point already in the reactions that occur upon placing a piece of zinc metal in a solution of copper(II) sulfate. In an electrochemical cell many processes occur simultaneously at all interfaces, including heterogeneous electron transfer (i.e. electrons hoping from the electrode to species in solution and back again), solvent reorganization, electrophoresis, and so on. We, however, only observe a net consequence of all the reactions—that is, we measure a voltage or a current, for example. Also, since electron transfer from the electrode only occurs at the electrode, the overall rate of a process depends on how rapidly species move to the electrode and away from it; so consider mass transport (diffusion, convection, migration, etc.).

iii) Although it is easy to overlook, the current measured is a direct manifestation of the overall rate of reaction(s) (i.e. the net rate of oxidation/reduction) and is identical at both (working and counter) electrodes.

[1] Bard, A. J., and Faulkner, L. R., 2001, *Electrochemical Methods: Fundamentals and Applications, 2nd ed.*, New York: Wiley Global Education.

[2] *Understanding Electrochemistry: Some Distinctive Concepts*, L. R. Faulkner, *J. Chem Ed.* 1983, 60, 262.

[3] This statement should be caveated, as in reactions with mediators and electrochemical mechanisms where reactions that are important occur away from the electrode. However, even in these cases the distance is of the order of a few tens of μm—within the Nernst diffusion layer.

iv) Similarly, potential (voltage, EMF) is an expression of the energy of an electron. Experimentally, however, we can only measure differences in potential; that is to say, in rather loose terms, the difference in energy of the electrons at each of the electrodes.

v) You can control potential and measure current (voltammetry) or you can control current and measure voltage (Galvanometry), but you cannot control both current and voltage simultaneously!

In summary, when interpreting data obtained from electrochemical experiments we should maintain awareness that what we measure (current/potential) is dependent on the nature of the chemical processes that occur, but also on the state of the electrode solution interface, mass transport, and the limits of the electronic components in the entire system. This awareness is especially important when we develop models to fit experimental data; a good model is not just a model that gives a good fit to the data but is one that makes chemical and physical sense and is robust to the outcome of further experiments.

Sibrandus Stratingh (1795–1841) by Johan Joeke Gabriel van Wicheren c. 1839. University Museum Groningen

This book would not be possible without the many undergraduates and graduates (and especially my research group) who have encouraged me over the years by their enthusiasm and critical feedback. I cannot name them all but there are students, Tim Meinds and Yvonne Halpin, and colleagues, Johan Hjelm, Ronald Hage, and Jaap de Jong who have made the most impact on driving my enthusiasm for electrochemistry. And of course, the first electric car built in 1835 by Professor S. Stratingh, the namesake of the Stratingh Institute for Chemistry.

As a final remark, my partners on the journey this book has been on should not go unmentioned. My sister, Nicola, and my editor, Martha Bailes, both for their constant encouragement and for life's experiences we have shared this last year.

Replica by Anton Stoelwinder, Gorredijk, The Netherlands

Wesley Browne
Stratingh Institute for Chemistry
University of Groningen
2017

Table of units and symbols

Quantity/Symbol		Unit(s)	
E_A	activation energy	$J\ mol^{-1}$	joules per mole
ω	angular frequency	$rad\ s^{-1}$	radians per second
A	area	m^2	square metre
k_B	Boltzmann coefficient	$J\ mol^{-1}\ K^{-1}$	joules per mole per Kelvin
Q	charge	C	Coulomb
C	capacitance	F	farad
χ_c	capacitive reactance	Ω	ohm
C_{dl}	double layer capacitance	F	farad
ε	relative permittivity (dielectric constant)		dimensionless ratio
σ	conductivity	$S\ m^{-1}$	siemens per metre
I	current	A	ampere
i	current density	$A\ m^{-2}$	ampere per square metre
ε	efficiency		
H	enthalpy	$J\ mol^{-1}$	joules per mole
ΔH	change in enthalpy	$J\ mol^{-1}$	joules per mole
$\Delta_R H^o$	change in enthalpy under standard conditions	$J\ mol^{-1}$	joules per mole
$\Delta_R H^\ddagger$	enthalpy of activation	$J\ mol^{-1}$	joules per mole
S	entropy	$J\ mol^{-1}$	joules per mole per Kelvin
ΔS	change in entropy	$J\ mol^{-1}\ K^{-1}$	joules per mole per Kelvin
$\Delta_R S^o$	change in entropy under standard conditions	$J\ mol^{-1}\ K^{-1}$	joules per mole per Kelvin
$\Delta_R S^\ddagger$	entropy of activation	$J\ mol^{-1}\ K^{-1}$	joules per mole per Kelvin
EMF	electromotive force	V	volt
j_o	(exchange) current density	$C\ s^{-1}$	Coulombs per second
j_a	(anodic) current density	$C\ s^{-1}$	Coulombs per second
j_c	(cathodic) current density	$C\ s^{-1}$	Coulombs per second
z	charge number	(none)	
μ	chemical potential	J	joules
D	diffusion coefficient	$cm^2\ s^{-1}$	square centimetre per second
j	flux density	$mol\ m^{-2}\ s^{-1}$	mole per square metre per second
ϕ	Galvani potential	V	volts
G	Gibbs energy	$J\ mol^{-1}$	joules per mole
ΔG	change in Gibbs energy	$J\ mol^{-1}$	joules per mole

$\Delta_R G^o$	change in Gibbs energy under standard conditions	$J\ mol^{-1}$	joules per mole
$\Delta_R G^{\ddagger}$	Gibbs energy of activation	$J\ mol^{-1}$	joules per mole
J	imaginary number ($\sqrt{-1}$)		
Z	impedance	Ω	ohm
Z'	(real) impedance	Ω	ohm
Z_{Re}	(real) impedance	Ω	ohm
Z''	(imaginary) impedance	Ω	ohm
Z_{Im}	(imaginary) impedance	Ω	ohm
L	inductance		
E	permittivity	$F\ m^{-1}$	farad per metre
ϕ	phase shift	rad or ^{o}	radian or degree
ΔE	potential difference voltage	V	volt
η	overpotential	V	volt
R	resistance	Ω	ohm
λ	relative parameter for charge transfer		
κ	transmission coefficient		
Γ	surface concentration	$mol\ cm^{-2}$	moles per square centimetre
W_{elec}	electrical work	$J\ s^{-1}$	joules per second
W_{therm}	thermal work	$J\ s^{-1}$	joules per second
W	Warburg impedance	Ω	ohm
R	ideal gas constant	$8.314\ J\ mol^{-1}\ K^{-1}$	
F	Faraday constant	$96\ 489\ C\ mol^{-1}$	
k_B	Boltzmann constant	$1.380\ 10^{-23}\ J\ mol^{-1}\ K^{-1}$	
h	Planck constant	$6.626\ 10^{-34}\ J\ s$	

Note that

Ω F = s (second)

V C = J (joule)

OCP	open circuit potential
Q	reaction quotient
α	activity
γ	activity coefficient
iR_u	'iR' drop
R_u	uncompensated resistance
K_{sp}	solubility product

Table of equations

General (Chapters 1 to 3)

$V = IR$

$P = VI = I^2R = V^2/R$

$$R = \frac{(I \times L)}{A}$$

$\varepsilon = I\,(R + r)$ or $V = -r\,I + \varepsilon$

$Q = nF$

$$Q = -C\Delta E = -\left(\frac{\varepsilon A}{L}\right)\Delta E$$

$$I = \frac{dQ}{dt} = C\frac{dE}{dt}$$

$E_{applied} - iR_{sol} = E_{true}$

$$E_{true} = E_{applied}\left[1 - \exp\left(\frac{-t}{R_u C_{dl}}\right)\right]$$

$$E_{overall} = \frac{(n_1E_1 + n_2E_2 + n_3E_3 + \ldots n_NE_N)}{(n_1 + n_2 + n_3 \ldots\ldots n_N)}$$

Mass transport relations

Distance traveled in one dimension by diffusion $x = (2D_o t)^{½}$

$J = -D\left(\frac{dC}{dx}\right)$ Fick's 1st law

$\frac{dC}{dt} = D\left(\frac{\partial^2 C}{\partial x^2}\right)$ Fick's 2nd law

$i_j = \frac{z_j^2 F^2 A D_j C_j}{RT} \cdot \frac{\partial \phi}{\partial x}$ (current due to motion driven by an electric field)

Thermodynamic relations

$\left(\frac{\partial G}{\partial n_i}\right)_{T,n_j} = \mu_i$ chemical potential

$\left(\frac{\partial \Delta G}{\partial T}\right)_P = \Delta S$ entropy change

$\bar{\mu} = \mu + zF\phi$ electrochemical potential

$\Delta G = \Delta H - T\Delta S$ Gibbs energy relation

$\Delta G^o = -nFE^o = -RT \ln K_{eq}$

$$E = \frac{dE}{dT}T - \frac{\Delta H}{nF}$$

$w_{elec} = -\Delta G = -(-\Delta H - T\Delta S) = -\Delta H\,(1 - T\Delta S/\Delta H)$

$w_{thermal} = q_{total}\varepsilon = -\Delta H\,(T_h - T_c)/T_h$

$\Delta G = \Delta G^o - RT\,lnQ$

$E = E^o - \frac{RT}{nF}\,lnQ$ Nernst equation

$E_j = (t_+ - t_-)(RT/F)ln\,(\alpha^{RHS}/\alpha^{LHS})$ junction potential

Electron transfer kinetics (Chapter 4)

$j = j_a - j_c = \nu Fk_a\,[R] - \nu Fk_c\,[O]$

$k = B_c\,\exp(-\Delta G^{\dagger}/k_BT)$ where $B_c = (\kappa k_B T/h)$

$k_{red} = k^o\,\exp\,[-\alpha_c nF(E - E_c^{o\prime})/RT]$

$k_{ox} = k^o\,\exp\,[-\alpha_a nF\,(E - E_c^{o\prime})/RT]$

$j = vFB_a[R]\exp\left[-\frac{\Delta_{ox}G^{\dagger}}{RT}\right] - vFB_c[O]\exp\left[-\frac{\Delta_{red}G^{\dagger}}{RT}\right]$ Butler–Volmer formulation

$\eta = E_{applied} - E_{½}$ overpotential

$$j = j_a - j_c = j_o\exp\frac{(1-\alpha)F\eta}{RT} - j_o\exp\frac{(-\alpha)F\eta}{RT}$$

$i = a'\exp(b'\eta)$ or $\eta = a + b\log_{10} i$ relation between current and overpotential

Voltammetry (Chapter 5)

$I = nFA\frac{d\Gamma_{ox}}{dt} = -nFA\frac{d\Gamma_{red}}{dt}$ for surface confined process

$$I = nFAD_A\left(\frac{d[Red]}{dx}\right)_{x=0}$$

Cottrell equation: $i_d(t) = nFAD^{½}C_o/\pi^{½}\,t^{½}$

Diffusion limited current: $I_{lim} = nFA\left(\frac{D_{Red}}{\delta}\right)[Red]_o$

Catalytic current: $I = nFAC\sqrt{Dk'_{cat}}$ and hence $I \propto \sqrt{k'_{cat}}$

For a reversible electrochemical reaction

$I_p = (0.4463)\,nFA\,(nFvD/RT)^{½}C$ under diffusion controlled conditions at a planar electrode

$I_p = (2.69.10^5)\,n^{3/2}AD^{½}v^{½}\,C$ under diffusion controlled conditions at a planar electrode at 298 K

$I_p = \frac{n^2F^2}{4RT}vA\Gamma^*_o$ for a surface confined process

$I_p = (9.39\times10^5)n^2vA\Gamma^*_o$ for a surface confined process at 298 K

For an irreversible electrochemical reaction at 298 K (under diffusion control)

$$I_p^{ox} = (2.99 \times 10^5)\, n(\alpha n_a)^{1/2} A D_A^{1/2} v^{1/2} C$$

Limiting current at a microelectrode or under hydrodynamic control

$I_{lim} = 4nFD_A r_e C_o$ (r_e is the radius of the electrode)

Electrochemical impedance spectroscopy (Chapter 6)

$$|Z| = \sqrt{Z_{Re}^2 + Z_{Im}^2}$$

$$E_t = E_{max} \sin\omega t \quad I_t = I_{max} \sin(\omega t + \phi)$$

$$Z(\omega) = \frac{E(t)}{I(t)} = \frac{E_{max} \sin(\omega t)}{I_{max} \sin(\omega t + \phi)} = |Z| \frac{\sin(\omega t)}{\sin(\omega t + \phi)}$$

$$Z = Z_{Re} + j\, Z_{im} = |Z|(\cos\phi + j \sin\phi)$$

$$Z^2 = (Z_{Re} + j\, Z_{im})\,(Z_{Re} - j\, Z_{im}) = Z_{Re}^2 + Z_{im}^2$$

$$\phi = \tan^{-1} \frac{|Z_{Im}|}{|Z_{Re}|}$$

Impedance of a resistor: $Z(\omega) = R$

Impedance of a capacitor: $|Z(\omega)| = \sqrt{(-\, j/\omega C)^2}$

$Z^* = Z_1 + Z_2 + Z_3$......when in series and $\frac{1}{Z^*} = \frac{1}{Z_1} + \frac{1}{Z_2} + \frac{1}{Z_3}$.... when in parallel

Introduction to electrochemistry

1.1 Introduction

Electrochemistry is sometimes seen as 'all about the electrode' but, in reality, an entire system needs to be considered, including the solvent, electrolyte, membranes, wires, and connections, to name but a few components. In this chapter, we will discuss several of the basic concepts and ideas pertinent to electrochemistry beginning with the concept of charge moving in a circuit and spontaneous polarization. We will then move on to discuss electrical components that are familiar to us in everyday life, even if we do not usually give much thought to them. As we will see in later chapters, 'solid state' concepts such as resistance, capacitance, and potential difference are also central to understanding solution electrochemistry.

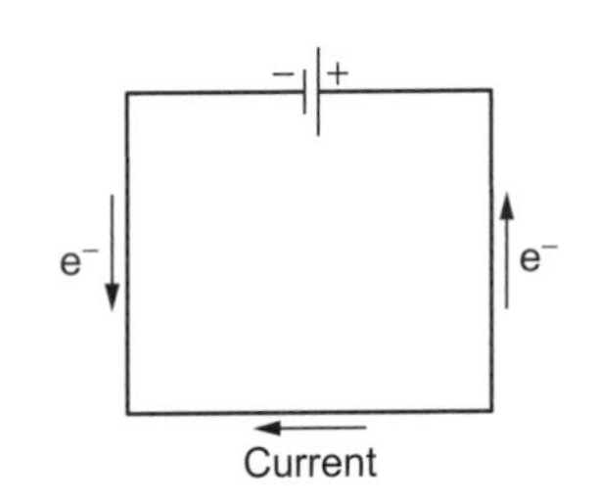

Figure 1.1 Electrons move in the opposite direction to current in a wire circuit.

1.1.1 Movement of charge

Electrochemistry is the movement of charge (Fig. 1.1), including the movement of electrons in a wire to or from an interface—for example, capacitors, membranes, and so on—and the movement (migration) of ions in a liquid or solid solution, and, above all, the movement of electrons from a solid to a species in solution and vice versa. When the movement of charge is at a liquid/solid interface (Fig. 1.2), the solid conductor is referred to as an **electrode**; that is, either an anode if the net movement of electrons is from species in the liquid phase to the solid or a cathode if from the solid to species in the liquid phase.

The flow of charge (Q), as 'electric current' (I), is dependent on the ability of a material to allow charge to move; that is, how conductive the material is. If current flows through a solution then it is referred to as a conductor, although the conduction is through the movement of ions and not electrons (Fig. 1.3). Examples of (i) electron conductors are metals, graphite, doped semiconductor metal oxides, PbO_2, polyaniline, and (ii) ionic conductors are sea water (Na^+(aq), Cl^-(aq), etc.), ZrO_2 (O_2^-), AgCl (Ag^+).

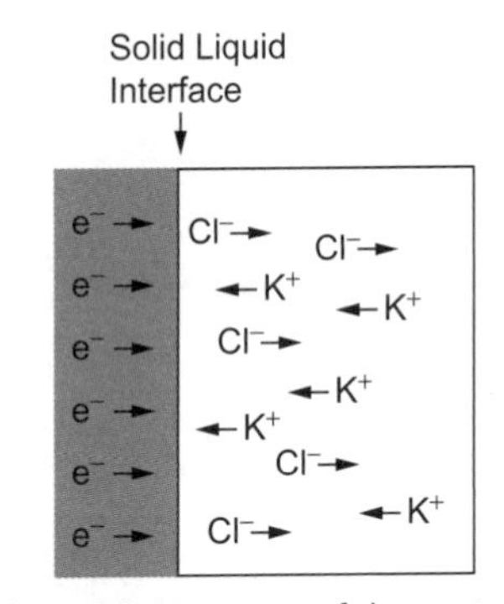

Figure 1.2 Movement of charge at (but not across) a metal/electrolyte interface (capacitive current).

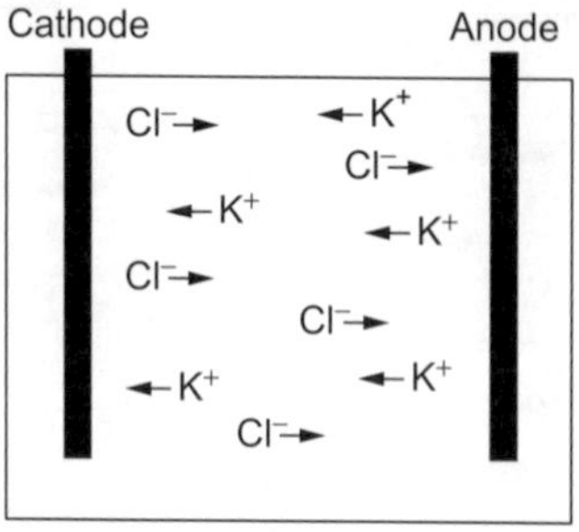

Figure 1.3 Ions moving in a solution between electrodes (see also Chapter 3 for transfer numbers).

Tension is another term used for voltage, especially when discussing power lines.

Box 1.1

A perfect insulator does not conduct electric current; that is, static electricity obtained by rubbing different non-conducting materials together is retained at the point where charge separation occurs until sufficient charge (and potential) is built up for a spark to develop.

Since the movement of charge (electrons and ions) requires energy (more precisely an energy or potential gradient), a force is needed to induce the movement. This force is referred to, depending on the situation, as electromotive force (EMF), potential, cell voltage, cell potential—and the symbols used (often loosely) to express it are E, E^o, $E_{½}$, E_{cell}, and V—depending on context. Although each term and its symbol relates to a specific concept/situation, they are all related by the general concept of electrochemical gradients and energy differences. If the movement of charge in a material does not involve the loss of energy, then the material is referred to as a superconductor, which is familiar to most chemists that have used a modern NMR spectrometer in which the magnetic field is provided by a coil of superconducting wire. However, even in the most conductive materials encountered in everyday life, some energy is lost as heat (resistive heating) as current flows. The less energy that is lost, the more conducting the material is. Indeed, almost every material falls between being a superconductor and an insulator, that is, metallic, semiconductor, and so on; where a particular material falls on this scale determines how much energy is lost as **resistive heating**.

In a Galvanic cell the reactions at both electrodes drive current through the circuit spontaneously. In an electrolytic cell, a power supply is added at the point close to where the ammeter (A) is located.

In electrochemistry, (electric) current is carried in any of several ways around a circuit. The diagram in Fig. 1.4 illustrates this movement in a Galvanic cell. The electrons flow from the anode (where net oxidation takes place) to the cathode (where net reduction takes place) via the wire (and through the ammeter and resistor). Electrons cannot flow through the solution between the two electrodes. Instead charge is carried by the movement of ions as indicated by arrows.

There is a problem though; in order for current to flow, we need to create a potential difference (i.e. an EMF). In addition, we cannot indefinitely remove electrons from the anode and deliver them to the cathode since as the charge on each electrode increases so too does the difficulty in removing or adding further electrons and at a certain point the electrodes would be coated with a full layer of positive (or negative) ions. The process of creating net positive or negative charge at an electrode surface is called polarization; we need to supply electrons to the anode and remove electrons from the cathode in order to depolarize the electrodes. The electrons are supplied and removed from the electrode (depolarization) by redox reactions with species in solution.

At the anode side, species can give up electrons (i.e. they are oxidized). Some examples are shown below the anode in Fig. 1.4. Electrons are removed from the cathode by reducible chemical species (some examples are given below the cathode). A basic requirement is that there must be a net (overall) driving force (ΔG is negative) for the anode and cathode reactions; that is, the cell reaction overall must be spontaneous. If ΔG is negative, then the cell is called a **Galvanic cell**, if ΔG is positive but reactions take place by applying an external voltage (i.e. putting energy into the system) then it is referred to as an **electrolytic cell**. We will go deeper into Galvanic and electrolytic cells in Chapter 2.

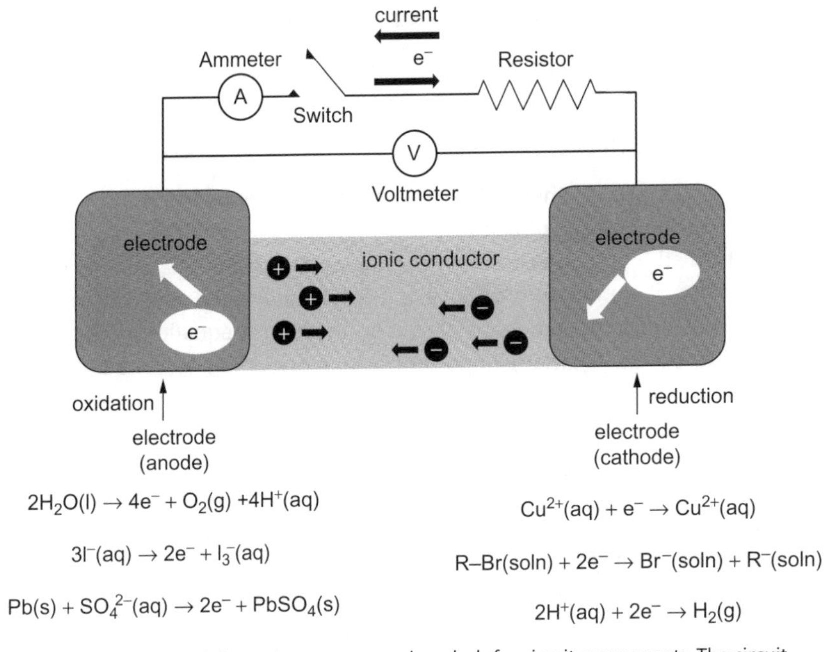

Figure 1.4 Summary of electrode processes and symbols for circuit components. The circuit shown includes a switch (the line with triangles at the end), a load (usually a resistor represented by the zig/zag line), an ammeter for measuring current flow (A) and a voltmeter for measuring the potential difference (V) between two points in the circuit (i.e. across the two electrodes).

1.1.1.1 The Coulomb (C), charge (Q), and current (I)

In electrochemistry, for historical reasons, charge (with the symbol Q) is not expressed in terms of the number of ions or electrons but instead the SI unit 'the Coulomb' is used (unit symbol C). The charge of one electron is 1.6022×10^{-19} C (known as the elementary charge). This means that 1 mole of electrons has a total charge of 96 485 C; that is, the Faraday constant. The Faraday constant is used to relate charge (in C) and the number of moles of electrons by the equation: $Q = nF$. The origin of this relation, Faraday's laws, rests in the observation that electrochemical reactions (at least some reactions) take place with the loss/gain in mass at an anode/cathode in direct proportion to the amount of direct current passed. The current flowing in a circuit is the amount of charge (Q) that passes a certain point per unit time and mathematically this is expressed as $I = dQ/dt$.

Faraday's 1st Law of Electrolysis—The change in mass of an electrode during electrolysis is directly proportional to the charge passed at that electrode.

Faraday's 2nd Law of Electrolysis—the mass change of an electrode composed of a single element for a given amount of charge passed by direct current is directly proportional to the element's mass.

1.1.1.2 Confusing conventions

The anode and cathode shown in Fig. 1.4 become the cathode and anode, respectively, in an electrolytic cell. This inversion is a common source of confusion and will be discussed later. An easy way to avoid confusion is to remember that oxidation of species present in solution occurs at the anode and reduction of species occurs at the cathode. In addition, it is convention that the cathode is drawn on the right and the anode is drawn on the left-hand side with assumption that the cell is a Galvanic cell.

1.1.2 **Terminology in oxidation and reduction**

From an electrochemical perspective, we are mostly concerned with gain and loss of electrons. An example is the reaction:

$$Fe^{3+} + V^{2+} \rightleftharpoons Fe^{2+} + V^{3+}$$

In this reaction, Fe^{3+} is the oxidizing agent (oxidant) while V^{2+} is the reducing agent (reductant) when going from left to right. It is important to realize that at an electrode the same species is undergoing oxidation and reduction continuously and what we measure/observe is the net result of an equilibrium being established—hence the emphasis placed on the Nernst equation in this book. This point will be especially important in our discussion of voltammetry in Chapter 5.

Oxidation and reduction mean different things in different situations:

Oxidized	Reduced
oxidant	reductant
loss of electrons	gain of electrons
anode	cathode
removal of hydrogen	removal of oxygen
addition of oxygen	addition of hydrogen

1.1.3 **Chemical oxidation and reduction**

Electrochemistry is primarily concerned with the exchange of electrons. Exchange between species in solution is (relatively) simple, however, electrons can be exchanged between species in solution and species in the solid state (i.e. an electrode). A simple example of this phenomenon is the changes that occur when a piece of zinc metal, which has been polished to remove the metal oxide layer, is dipped into an aqueous solution of copper(II) sulfate. The zinc metal quickly becomes coloured by a layer of metallic copper deposited from solution. If we do the reverse; that is, take a strip of polished copper and dip it into a solution of zinc(II) sulfate, nothing will appear to happen.

In fact, some chemistry will actually happen but it will be more or less the same as when Zn^{2+} ions were not present.

So, what actually happens and why does it happen?

1.1.4 **Polarization and matching chemical potentials**

The overall reaction

$$Zn(s) + Cu^{2+}(aq) \rightleftharpoons Zn^{2+}(aq) + Cu(s)$$

is spontaneous as written (as we will discuss in Chapter 2).

However, it can be thought of, and indeed is, a combination of two entirely separate reactions (so called half-reactions, Fig. 1.5)

$Cu^{2+}(aq) + 2e^- \rightarrow Cu(s)$	The Cu^{2+} ions receive two electrons from the zinc strip (electrode).
$Zn(s) \rightarrow Zn^{2+}(aq) + 2e^-$	The zinc strip releases Zn^{2+} ions into solution to obtain the electrons needed to reduce the Cu^{2+} ions.

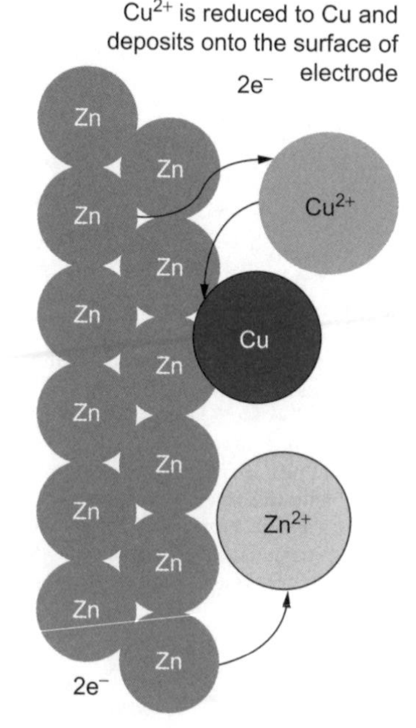

Figure 1.5 Processes occurring when a piece of zinc is placed in a solution containing Cu(II) ions.

However, which reaction happens first? Or do both reactions have to occur simultaneously? When a metal is placed in a polar solution a phenomenon called polarization occurs. The metal has a chemical potential that is different to that of the solution. It is not possible to maintain a sudden change in chemical potential at the interface—instead they tend to equalize. The metal gives up metal ions into the solution (i.e. Zn(II) ions from Zn) and becomes negatively charged. The Zn(II) ions in solution stay, for the most part, close to the electrode. Hence, the second of the two half-reactions shown here actually takes place whether or not the Cu(II) ions are present in solution.

The symbol ⇔ signifies a reversible reaction in this document, however, for convenience half-cell reactions are often written with a → despite representing microscopically reversible reactions.

The difference made by the Cu(II) ions is that they can take up electrons from the now negatively polarized electrode to form metallic Cu(0); a process that is called depolarization (Fig. 1.6). The electrode, however, needs to re-equalize its chemical potential with the solution and does this by releasing yet more Zn^{2+} ions. Both processes continue and drive the oxidation of the Zn(0) to Zn(II) ions well beyond the equilibrium position that it would hold if the Cu(II) ions were not present. Eventually the surface of the zinc strip will be coated with a layer of copper atoms. At this point the electrode is de facto a copper electrode and the only process that will occur is the continued exchange of Cu atoms and Cu^{2+} ions at the electrode. The process overall is called electroless deposition or electroless plating.

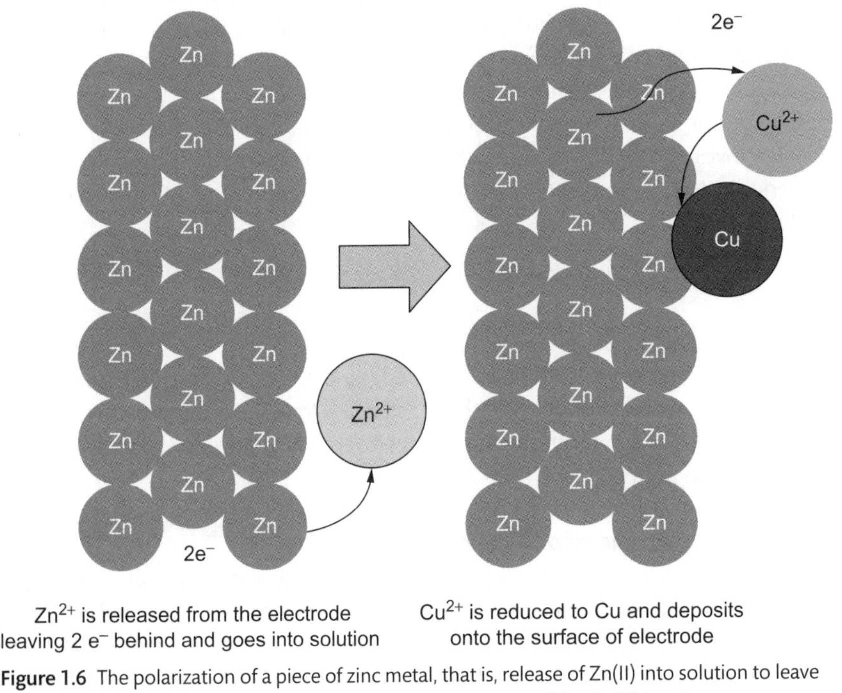

Figure 1.6 The polarization of a piece of zinc metal, that is, release of Zn(II) into solution to leave a negatively charged piece of metal occurs even in the absence of the Cu(II) ions but stops once the chemical potential of the metal matches that of the solution. If Cu(II) ions are present, they depolarize the electrode by taking up the electrons and depositing on the electrode forcing the release of Zn(II) ions into solution to continue.

1.2 Basic concepts in electrical circuits

Applying electrochemistry to chemical problems requires a basic understanding of electronics; after all, the electrochemical cell is mostly wires and circuit boards with the sample of interest being only one of many components. In this section, a number of important concepts and phenomena in electronic circuitry will be introduced and explained briefly.

1.2.1 Alternating current and direct current (AC/DC)

When current flows, electrons move along a conducting wire and the metric unit of current is the ampere (A). In a way, the flow of electrons (current) can be imagined as water (the electrons) flowing through a pipe—however, one should remember that the electrons move along the surface of a wire and not down the middle. Electrons move in a circuit down a gradient called a **potential difference**; that is, from a point of higher potential energy to a point of lower potential energy. The potential difference is quantified in volts (V) and can be considered as being analogous to water pressure in a pipe. Curiously, the movement of electrons is in the opposite direction to the direction of the current (this is an historical anomaly due to the fact that electricity and current was discovered long before the electron was discovered). Current 'flows' from positive to negative poles and electrons 'flow' from negative to positive poles.

The electroneutrality principle states that significant net charge cannot reside in any macroscopic volume within a conductor; that is, positive or negative charge cannot accumulate or 'flow through' the conductor. Like charges repel each other and move as far from each other as possible—by residing on the surface of the metal; the Faraday effect. Faraday cages are important for electrical safety and metallic suits that cover the full body are used by workers when repairing live high-tension power cables.

Current passes as either AC or DC (Fig. 1.7). Direct current (DC) means that the current flows in one direction all the time. Alternating current (AC), by contrast, involves the flow of current in which the direction and strength changes periodically (sinusoidally). The current can be a combination of DC and AC also (Fig. 1.7).

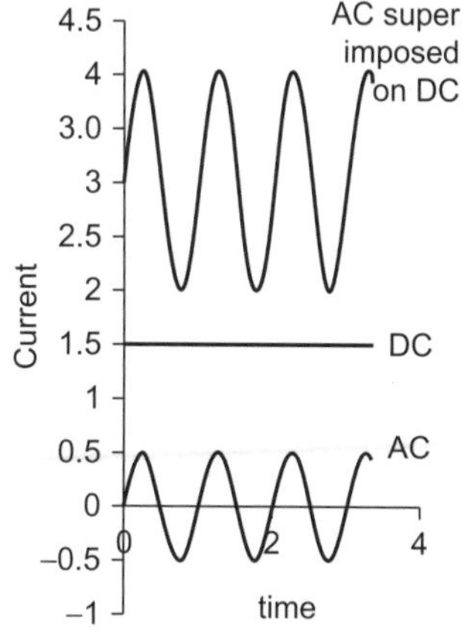

Figure 1.7 Current as a function of time for AC (alternating current) and DC (direct current).

1.2.2 Ohm's law and resistivity

The flow of electrons is impeded (resisted) by the material used to make a wire. For gold, silver, and copper wire the resistance is lowest. For other materials such as tungsten (used in incandescent light bulbs) the resistance to the movement of electrons is greater.

Resistance is measured in ohms (Ω) and the resistance of a material is dependent on its length and cross-sectional area by

$R = (I \times L)/A$ where I is the resistivity, L is the length of the wire, and A is the cross-sectional area of the wire. Note the dimensional analysis, $R(\Omega) = (I(\Omega m) \times L(m)) / A(m^2)$

The resistance of a material of unit length is referred to as its resistivity (I) and is expressed in ohm metres (Ω m), which is independent of the conductor's other dimensions (i.e. how thick or thin the wire is).

Copper has a resistivity of $1.72 \times 10^{-8}\ \Omega$ m. A 1 m long copper wire with a cross-sectional area of $1 \times 10^{-6}\ m^2$ will have a resistance of:

$$R = (1.72 \times 10^{-8}\ \Omega\ m \times 1\ m)/1 \times 10^{-6}\ m^2 = 0.017\ \Omega$$

Ohm's law states that direct current (DC) is proportional to potential difference divided by the resistance, or in other words the current that flows is directly proportional to the applied voltage and the constant of proportionality is $1/R$.

$$I = V/R\ \{1\ A = 1\ V/1\ \Omega\}$$

The 'voltage drop' along 1 m of the copper wire when 1 amp of current is flowing will be:

$$V = IR = 0.017\ \Omega \times 1\ \text{A} = 17\ \text{mV}$$

A simple circuit (Fig. 1.8) uses a battery to provide the direct current that passes through a resistor. The current that flows in the circuit is given by $I = V/R = 3\ \text{V}/100\ \Omega = 30\ \text{mA}$.

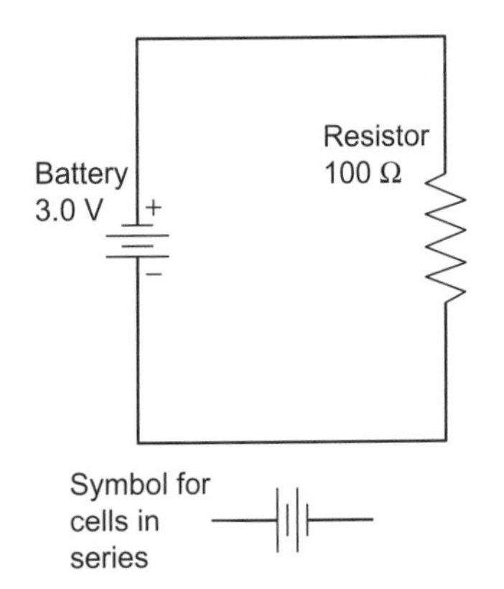

Figure 1.8 Simple circuit.

1.2.3 **Resistance to direct and alternating current**

When electricity flows through an electrical circuit continuously in one direction (direct current) the energy lost (determined by the voltage drop across the resistor) is related to the **impedance** of the circuit and its components. When the direction and current change sinusoidally, that is, AC, the resistance to the current is referred to as impedance.

1.2.3.1 **Impedance**

The term impedance in modern English is the ability of something to impede (restrict the progress of) something else; that is, reduce the ability of the latter to move in a particular direction. A simple example is the narrowing of a pipe at a certain point in a water system. The constriction will impede the flow of a fluid through that system. In electrochemistry the term impedance refers to the ability of a material to resist the movement of charge. We will meet impedance later, in detail, in the technique electrochemical impedance spectroscopy (EIS) in Chapter 6.

1.2.4 **Filaments in incandescent bulbs and resistive heating**

In old 'non-energy saver' light bulbs, light was generated by placing a filament of tungsten in a high vacuum and passing a current through it (e.g. 100 W at 240 V). The passage of current is resisted (impeded) and energy is lost as heat. This heat generation is referred to as resistive (also known as ohmic or Joule) heating. James Joule, in the 1840s, formulated what is now known as Joule's law while studying the effect of a change in the length of a wire and the current passed through it on the change in the temperature of the water the wire was immersed in.

Joule's first law. Power (P) = Voltage (V) by Current (I) ($P = VI = I^2R = V^2/R$)

Therefore:

$$100\ \text{W} = 240\ V \times I$$

$$100\ \text{W}/240\ \text{V} = I = 0.42\ \text{A}$$

$$V = IR \Rightarrow 240\ \text{V} = 0.42\ \text{A} \times R \Rightarrow R = 570\ \Omega$$

How thick does a 10 cm long tungsten filament need to be to have a resistance of 570 Ω?

$$\text{R} = \text{I} \times \text{L/A} \Rightarrow 570 = 5.4 \times 10^{-8}\ \Omega\ \text{m} \times 0.1\ \text{m/A} \rightarrow \text{A} = 9.47 \times 10^{-12}\ \text{m}^2$$

The ground or 0 V is in effect a point on the circuit chosen arbitrarily and is given the earth symbol. This may seem a bit uncomfortable but we take the same approach, arbitrarily picking a 'zero' or reference point, also when we discuss reduction potentials in Chapter 2.

Therefore, the diameter of the wire needs to be 3.4 μm; that is, the tungsten wire needs to be very thin to have sufficient resistance.

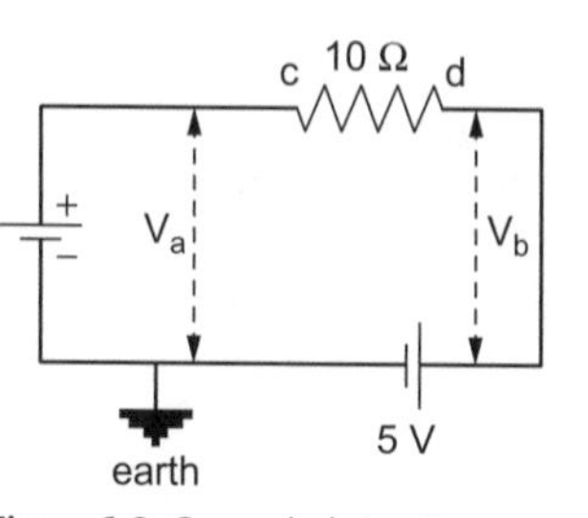

Figure 1.9 Grounded circuits.

1.2.5 **Measuring potential difference**

Electrical potential (V) at a point in a circuit is always measured relative to some reference point. The reference point is usually the earth (i.e. the connection to the ground).

The potential difference between two points, for example a and b (Fig. 1.9), is $V = V_a - V_b$. By convention current flows from high (positive) potential to low (negative) potential and electrons in the opposite direction. Point 'a' on the circuit is closer to the positive terminus of the battery with respect to the ground. The negative terminal of the battery is at the ground potential (0 V). For example, in the circuit shown in Fig. 1.9 the potential difference between point c and d (Fig. 1.9) is 5 V and the current is $V = IR \rightarrow 5\text{ V} = 10\,\Omega \times I \rightarrow I = 0.5$ A.

1.2.6 **EMF, voltage, and internal resistance in batteries**

EMF is the energy provided by a battery to a circuit and is the potential difference across the termini when current is not drawn (expressed in V). One point that is often overlooked is internal resistance in a battery. A battery pack generates, for example, an EMF (ε) of 3 V by combining two batteries that each provide 1.5 V in series. Consider again the diagram shown at the side. There are in fact more than three sources of resistance in the circuit. We will ignore the resistance of the wires and contacts as these are negligible under normal circumstances. The obvious source of resistance is the 100 Ω resistor (R). Two other sources of resistance in the circuit shown are the batteries themselves. Each of them exerts an internal resistance (that we will refer to as 'r'), which is usually small but nevertheless sufficient to affect the voltage across the resistor.

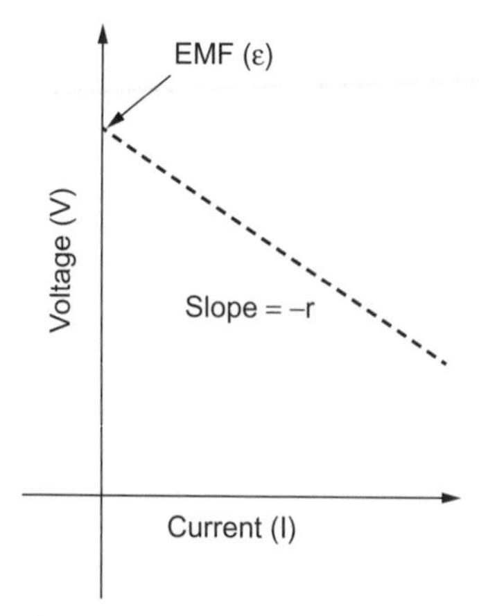

Figure 1.10 Dependence of cell voltage on current.

When current flows the internal resistance (r) will decrease the voltage across the resistor.

$$\varepsilon = I(R + r)$$

which when rearranged is $\varepsilon = I\,R + I\,r$ or $\varepsilon = V + I\,r$

where V is the potential difference measured between the battery terminals (and also across the 100 Ω resistor). The variables in the equation are I and V so if we rearrange this equation then we obtain an equation of a line where ($-r$) is the slope: $V = -r\,I + \varepsilon$

A graph of the dependence of voltage (measured across the terminals) on the current will give an intercept, which is the cell EMF (Fig. 1.10).

Internal resistance can present a problem in circuits that use a battery as a power source. When little or no current is flowing in a circuit then the system is essentially static and a potential difference between any two points is relatively straightforward to predict precisely. A problem arises when one wants to predict the voltage drop between two points in a system where a substantial current is flowing. The reason for this is that within the battery there is a solid or liquid solution containing electrolyte (see Chapter 2) that resists the flow of current also. However, unlike a conventional resistor, the resistance of the solution is not constant but instead can increase as the current increases (e.g. Warburg impedance) and as the battery is drained.

When current flows through a resistor, the potential energy of the electrons is converted to heat. This effect is useful, and indeed it is the basis of the incandescent lamp and glow bar heaters. In a battery, however, heating is a problem as it affects the energy available to move electrons (we will see in Chapter 3 that potential is temperature dependent). This topic will be dealt with later in Chapter 6; suffice to say that manufacturers make considerable efforts to minimize the internal resistance primarily because of the phenomenon referred to as resistive heating.

Warburg impedance can be thought of as the resistance to current flow because of mass transfer limitations: redox active species have to diffuse to the electrode and the rate of diffusion is dependent on concentration gradients. We will discuss this in more detail in later chapters.

In Chapter 6, we will return to the battery's internal resistance and it is worth noting that we will meet EMF again as **the open circuit potential (OCP)** of an electrochemical cell in Chapter 2.

$1\ F = 1\ C\ V^{-1}$

1.2.7 **Capacitance**

Capacitance is the movement of charge on either side of an interface but with no actual movement of charge across the interface, that is, neither electrons nor ions 'jump' across. When a capacitor (condenser) is charged, the metal contacts on each side of the capacitor are said to be polarized. The most commonly encountered example of capacitance is in capacitors of course (Fig. 1.11). A capacitor is in its simplest form comprised of two metal plates separated by a material with a high dielectric constant (i.e. an insulator that can respond to an electric field at a microscopic level but will not allow charge to actually flow through it, e.g. $CaCu_3Ti_4O_{12}$). The dielectric material can, however, be polarized; that is, dipoles can align in an electric field but charges cannot move. The capacitance (C in Farads, F) of a component in an electrical circuit is calculated from the product of the permittivity of the material separating the plates (ε in F m^{-1}) by the area of the facing plates (A in m^2) divided by the gap (L in m) and the potential difference across the capacitor. The amount of charge (Q) that accumulates at each plate can be calculated using the equation: $Q = -C\Delta E = -\left(\frac{\varepsilon A}{L}\right)\Delta E$

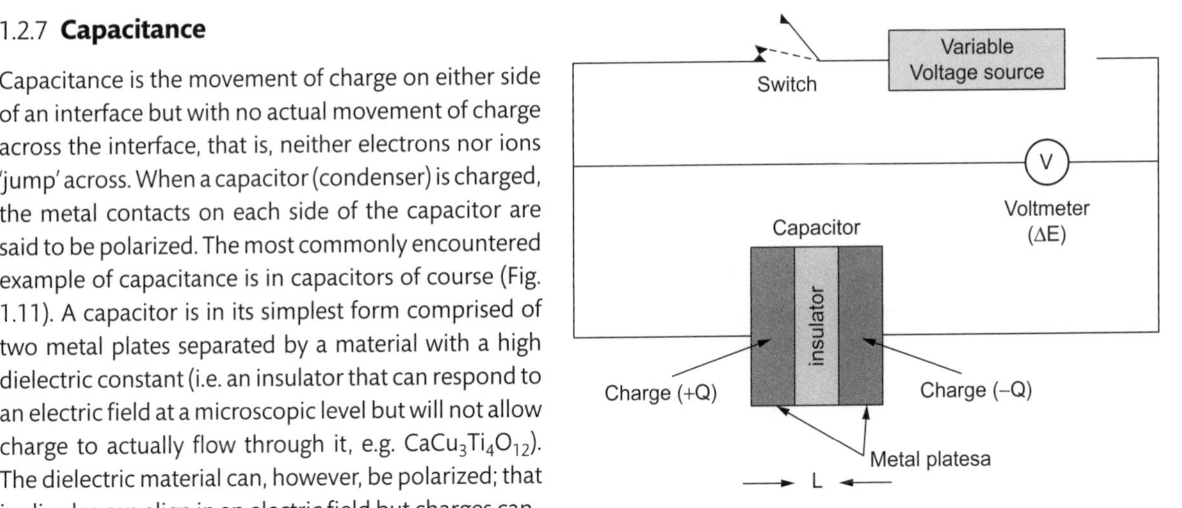

Figure 1.11 Circuit showing a capacitor in detail.

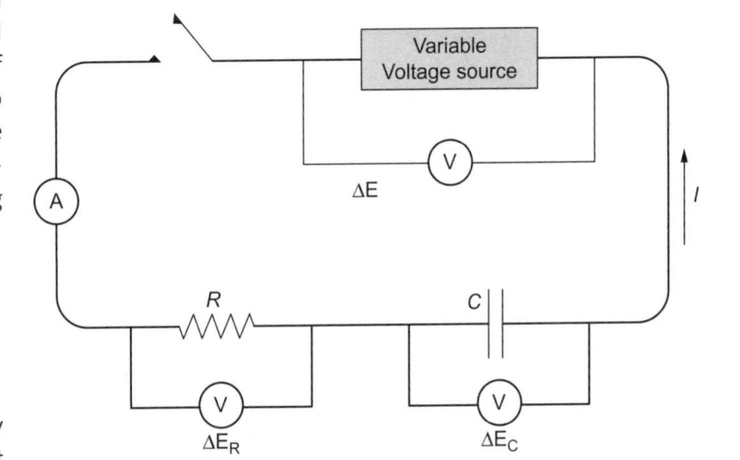

Figure 1.12 Circuit containing a resistor and capacitor in series.

1.2.8 **Time constant of an electronic circuit**

Although electronics operate at extremely high speeds, individual circuits (Fig. 1.12) that contain components such as resistors and capacitors have a finite time in which they respond to a change; for example, to the applied voltage. Indeed, the capacitance of an interface is important in electrochemistry as it limits the ability of an electrode solution interface to respond to a change in potential and it gives rise to an electrochemical cell's 'time constant', a characteristic that limits studies of fast processes.

The charge that can be built up in a capacitor Q is the product of the capacitance and the potential difference. The current that flows when charging is the rate of passage of charge (dQ/dt) (Fig. 1.13) and hence:

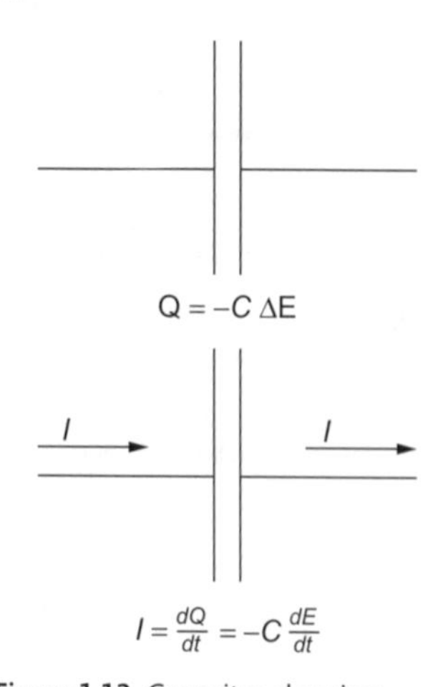

Figure 1.13 Capacitor charging.

As the capacitor is charged to its maximum storage, the current drops exponentially.

The potential difference across the capacitor is $\Delta E_c = -\Delta E_{source} - \Delta E_R$. The current that flows determines the voltage drop across the resistor; $I = -\frac{\Delta E_R}{R} = -C\frac{d\Delta E_c}{dt}$

and $\Delta E_R + \Delta E_C + \Delta E_{source} = 0$

The equation $I = -\frac{\Delta E_{source}}{R}\exp\left(\frac{-t}{RC}\right)$ can be derived from these two equations (Fig. 1.14).

The current can be expressed in terms of ΔE and an exponential term. The denominator is the so-called RC-time constant of a circuit. This will be important later when we consider how fast an electrochemical cell can respond to a change in applied voltage.

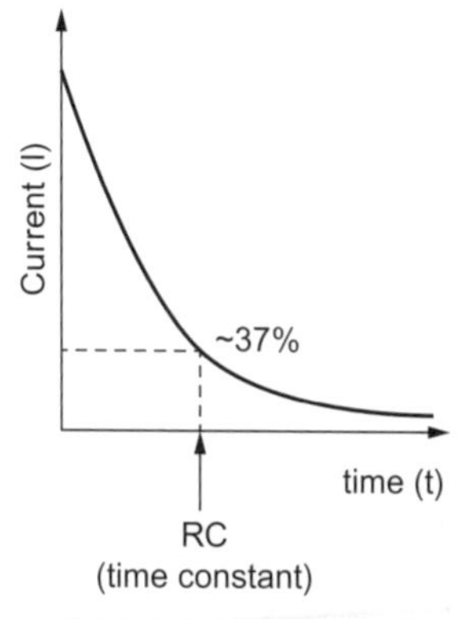

Figure 1.14 Current response of circuit to a change in potential (the increase in current is in fact slower than shown, see Chapter 5).

1.2.9 **Resistors and voltage sources in parallel and in series**

When we consider direct current through a circuit, it is relatively straightforward to calculate the potential difference and current flowing between any two points using Kirchhoff's laws (shown for resistors in series and in parallel in Fig. 1.15). When considering circuits, we should remember that (i) the potential difference between any two points is independent of the path followed and (ii) the current is the same at any point along a wire's length.

Resistors in series. When several resistors are arranged in series the total resistance across them is equal to the sum of their individual resistances. Remember that the current across each resistor is identical since electrons can't just disappear!

A loose analogy (and somewhat dangerous if taken too far!) is to think of electrons as water flowing through a pipe. One can imagine that in complex piping systems with connections the amount of water that flows in each pipe after a junction is dependent on the diameter of the pipes. The wider the pipe, the lower the resistance (impedance) to fluid flow and the greater portion of water that flows through it.

Resistors in parallel. When they are arranged in parallel the total resistance is calculated from the reciprocal values of each individual resistor. For resistors in parallel the voltage drop across each resistor is the same since it does not matter which way the electrons move from one point to another and hence the current passing each one will be different—think about the analogy with water in a pipe mentioned previously. For example, if we construct a circuit with three resistors R_1, R_2, R_3 in parallel, and a 5 V power supply (Fig. 1.16), then current will flow once the circuit is closed. Through which part of the circuit does the current actually passes and are there any losses? The resistors impede the flow of current, which means that after each resistor the driving force for moving electrons will have decreased but the current (number of electrons or charge per unit time) will be the same. At each junction the current can flow along any path; that is, through R_1, R_2, or R_3.

The sum of the current through each of the three resistors is equal the total current.

$$I_1 + I_2 + I_3 = I_{total}$$

The potential difference across each resistor is the same as the potential difference across the battery. However, the resistors impede the flow of current through them to different extents, with the 1000 Ω resistor impeding the flow of electrons the most in the case shown in Fig. 1.16.

Kirchhoff's first law states that the current flowing towards a point is equal to the current flowing away from that point. Hence, if a single wire is connected to two or more other wires then the current flowing in that wire is equal to the total net current flowing in all of the other wires combined.

$$I_{in} = I_{out1} + I_{out2} + I_{out3} + I_{out4} + \ldots$$

Kirchhoff's second (voltage) law states that the sum of all voltages in a closed loop is zero. In other words, in a circuit, the voltage drop across the entire loop is the same as the voltage delivered by the power source in that loop.

$$V_{total} = V_1 + V_2 + V_3 + V_4 \ldots \text{ and so on}$$

In addition, the voltage decrease (drop) over each branch is the same as the source voltage.

By using Ohm's law we can replace

$$I_{in} = I_{out1} + I_{out2} + I_{out3} + I_{out4} + \ldots$$

by

$$V/R_{in} = V/R_{out1} + V/R_{out2} + V/R_{out3} + V/R_{out4} + \ldots$$

the voltage terms cancel since they are all the same

$$1/R_{in} = 1/R_{out1} + 1/R_{out2} + 1/R_{out3} + 1/R_{out4} + \ldots$$

It is important to remember, however, that Kirchhoff's laws always hold for direct current systems and systems where changes occur slowly (remember the RC-time constant!). With AC the ability of individual components in the system to respond to changes in current/voltage can be too fast for the laws to apply at all times. This point will be discussed in more depth in Chapter 6, in the context of impedance spectroscopy. In the systems we will discuss in subsequent chapters, we will refer to the concepts described here repeatedly and describe components in electrochemical cells in terms of resistors and capacitors.

$$R_{total} = R_1 + R_1 + R_1$$

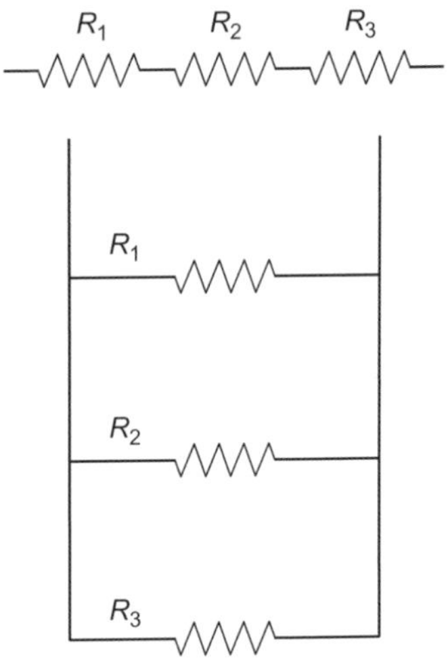

R1 R2 R3

R1

R2

R3

$$\frac{1}{R_{total}} = \frac{1}{R_1} + \frac{1}{R_2} + \frac{1}{R_3}$$

Figure 1.15 Resistors in series and in parallel.

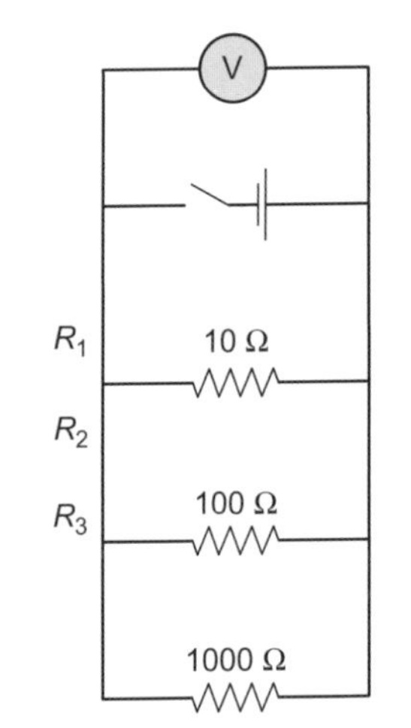

Figure 1.16 Resistors in parallel in a circuit.

1.3 Voltage, work, and Gibbs energy

An important relation in electrochemistry is between the potential difference (E) and Gibbs (free) energy (G). Gibbs energy is the amount of useful work that can be done by a chemical system (reaction) on its surroundings. We can relate electrical potential, which is the work that has to be done to move electrical charge from one point to another (measured in volts, V);

$$\text{Work} = \text{energy by charge} = E\,.\,q = V.\,C = J\ (\text{Joule})$$

The change in Gibbs energy, ΔG, under reversible conditions at constant temperature (isothermal) and pressure (isobaric) is the maximum possible electrical work done by a reaction on its surroundings.

Work done on the surroundings is equal to $-\Delta G$ (the Gibbs energy of a system decreases when work is done on its surroundings), hence

$$-\Delta G = \text{work} = q.E = nFE$$

which leads to an equation that is frequently encountered in electrochemistry $\Delta G = -nFE$

It is useful to reflect at this stage on the units for the terms in this equation:

$$\Delta G = -\,n\,F\,E$$

kJ mol^{-1} (dimensionless) (C mol^{-1}) (V)

The potential does not depend on the amount of the substance, nor does it depend on the number of electrons involved in the balanced equation for the reaction. Furthermore, 'n' does not have units and is simply a number reflecting stoichiometry. When we meet the Nernst equation in the next chapter we will reflect on this point again. A further relation that we will make use of is drawn by remembering that $\Delta G = \Delta H - T\Delta S$ and hence $-nFE = \Delta H - T\Delta S$. Therefore, we can use electrochemical measurements to determine thermodynamic properties. This point will be expanded upon in later chapters.

1.3.1 **Electrical current and reaction rates**

Although the aspects of electrochemistry covered in this chapter may appear unrelated to chemistry, in the following chapters we will see the close correspondence between current and voltage in solid state circuits, and current and voltage at electrode/solution interfaces and in solutions.

A final point to remember is that current is a rate ($I = dQ/dt$ with units of Coulombs per second) and the current flowing through an electrical circuit, in which a Galvanic or electrolytic cell is a component, is a manifestation of the rate of the reactions taking place at the solution electrode interfaces. What those reactions are and what we can learn about them by studying the voltage measured across the electrodes and the current that flows through the circuit will be the focus of Chapters 2 and 3, respectively.

1.4 Summary

This chapter should provide you with an understanding of:

- charge and the movement of charge (ions and electrons) as current
- polarization of electrodes and their depolarization

- a 'half-cell' reaction
- basic components of an electronic circuit; resistors, capacitors, electrodes, voltmeters, and ammeters
- Kirchhoff's laws and there use in calculating total impedance for a DC circuit
- the concepts of EMF, voltage, capacitance, impedance, and current, and their interrelations

1.5 Exercises

1.1. Describe the process of polarization of an electrode and the effect of a depolarizer.

1.2. A highly polished piece of copper is left to sit in a bath of water that has been deoxygenated and deionized before use. Describe the processes, if any, that you would expect to occur.

1.3. How thick does a 5 cm long tungsten filament need to be to have a resistance of 400 Ω?

1.4. Calculate the time constant for a circuit containing a 100 Ω resistor in series with a 15 μF capacitor and a 10 V power supply.

1.5. Calculate the total resistance in a circuit containing a 200 Ω resistor that is in series (a) with a pair of resistors, 100 Ω, and 250 Ω, in series and (b) with pair of resistors, 100 Ω and 250 Ω, that are in parallel with each other. If a 5 V battery is included in the circuit calculate the voltage drop across each of the resistors and the current flowing through each resistor.

2 The electrochemical cell

2.1 Introduction

In Chapter 1, we discussed electronic circuits with power supplies (e.g. a battery), resistors, capacitors, and switches, as well as what happens to a piece of zinc metal when placed in an aqueous solution containing Cu(II) ions. In this chapter, we will combine these topics to build an electrochemical cell and consider the various ways in which to represent and analyse electrochemical data (reduction potentials) in situations where there is no net change in conditions over time (i.e. potientiostatic conditions). Before doing so, first we need to consider the phenomenon of electrode polarization as well as the various classes of electrodes commonly encountered.

2.2 Electrode polarization and the relation between electrochemical cells and capacitors

The electrodes in this case are ideal polarizable electrodes. We will discuss polarizable and non-polarizable electrodes in Chapter 3.

The two **electrodes** in Fig. 2.1 are connected via wire with a switch. The position of the switch determines whether the electrodes are connected via a battery or a wire. When the switch is closed to connect the electrodes via the battery, the battery 'pushes' electrons to the end of one electrode and 'pulls' electrons from the other. However, each time an electron is moved from one electrode to the other it becomes more difficult to move the next electron. Eventually the negative and positive charge built up prevents more electrons from being pushed from one electrode to the other and the system reaches a steady state. The electrodes are now polarized; that is, they are no longer electroneutral. The air gap between the electrodes is a non-conducting low dielectric material and hence what we have made is a (quite poor) capacitor.

Addition of a solution containing an electrolyte (i.e. dissociated anions and cations such as aqueous KCl) to the box will allow more charge to build up at each electrode, as it has a higher dielectric constant than air. Furthermore, the positive and negative ions in the solution arrange at the electrode/solution interface to stabilize the charge built up. The net effect will be that there are slightly more negative ions on the positively polarized electrode and more positive ions at the negatively polarized electrode. We will discuss this in more

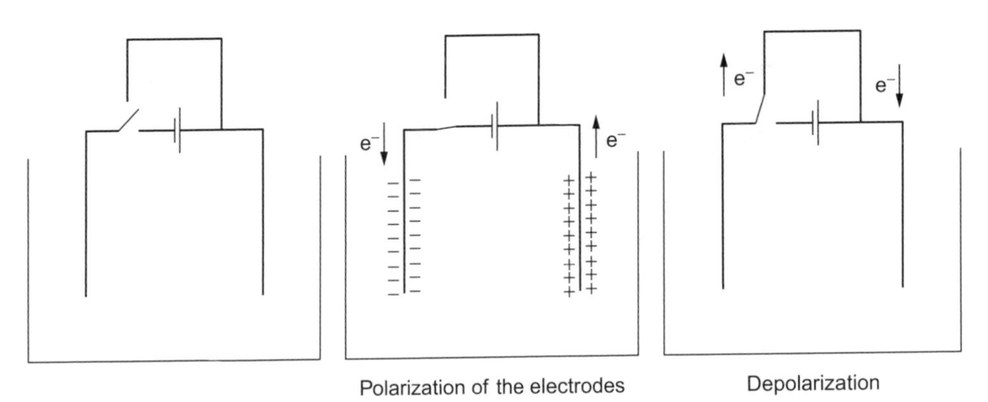

Figure 2.1 Closing the switch to connect the battery to both electrodes causes them to polarize. The electrodes depolarize when switched to bypass the battery and allow electrons to flow spontaneously from the negatively polarized electrode to the positively polarized electrode.

detail in Chapter 3. Once the electrodes are fully polarized then the situation will remain stable. Current will then flow only if the EMF generated by the battery is large enough to force electrons into solution (i.e. reduce species present) and to take electrons from species in solution (oxidation).

Any change in polarization will result in a change in the ordering of ions at the electrodes. The movement of ions through solution is much slower than the movement of electrons in a wire so the system will take a finite time to reach a new equilibrium situation. Hence, the flow of current measured, determined by the slowest process, will reflect the time it takes the ions to adjust to the change in potential (Fig. 2.2).

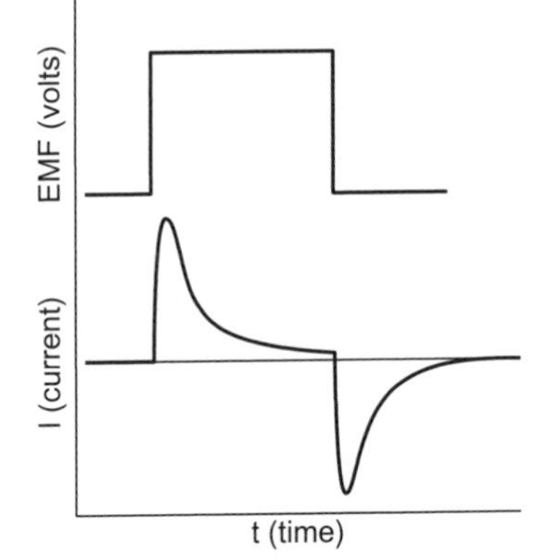

Figure 2.2 Current and voltage as a function of time when switched (see Figure 2.1) to connect the electrodes via the battery and afterwards to connect the electrodes via the wire.

Let's now consider what happens in the immediate time period that follows a change of the position of the switch to connect the electrodes via the wire. Notice that although the voltage changes abruptly, the current—that is, the rate at which electrons flow in the circuit—increases less quickly until it reaches a maximum and then falls to zero as the electrodes reach a new steady state of polarization. This delay in the current flowing following a change in potential is the response time of the circuit (or commonly quantified as the circuit RC time constant; see Chapter 1).

2.2.1 **Electrode classification**

At this point we are considering only electrodes as inert conducting materials, however, we cannot increase the potential between two wires (and hence their polarization state) in solution indefinitely. Eventually, a species in solution, on the surface of the electrode or even the electrode material itself, will begin to depolarize the electrode; that is, either something gives up or take up electrons to decrease the polarization of the electrode (this process is called depolarization). Electrodes are classified by the manner in which the electrode material and species in solution respond to a change in potential as being of the first, second, or third kind;

***Electrodes of the 1st kind**: the electrode is sensitive to the presence of ions of the metal the electrode is made of, that is* Metal/soluble metal ion such as Cu/Cu^+, Zn/Zn^{2+}.

***Electrodes of the 2nd kind**: where the metal ion forms concomitant with complexation to form an insoluble salt, that is* Metal/insoluble salt–$Pb/PbSO_4$, Hg/Hg_2Cl_2 or Ag/AgCl.

***Electrodes of the 3rd kind**: the electrode responds to a change in the redox state of a species that is not the same as the ion form of the metal from which the electrode is made.*

Metal/gas/ion–$Pt/H_2,H^+$

Metal/soluble ion, soluble ion–$Pt/Fe^{2+}, Fe^{3+}$

2.3 Galvanic (voltaic) cells

In dynamic electrochemistry, electrons are removed from, or given up to, species in solution and this process results in a change in energy. If energy has to be provided by an external power supply to drive the reactions then the electrochemical cell is referred to as an electrolytic cell. If the reactions occur spontaneously, as in the reaction of zinc metal with Cu(II) ions, then the cell is referred to as a Galvanic (or voltaic) cell. In this section, we will focus on Galvanic cells. In a Galvanic cell, a spontaneous reaction ($\Delta G < 0$) drives current through a circuit. In the reaction, one reagent undergoes (net) oxidation and the other (net) reduction.

Batteries and fuel cells are commonly encountered examples of Galvanic cells and will be discussed in Chapter 6.

In a Galvanic cell, the oxidation and reduction reactions are (spatially) separated. If we consider the reactions:

$AgCl(s) + e^- \rightarrow Ag(s) + Cl^-$ (in this reaction the silver ion is reduced to metallic silver)

$Cd^{2+} + 2e^- \rightarrow Cd(s)$ (in this reaction the Cd^{2+} ion is reduced to solid cadmium)

The balanced equation for the reaction is:

$$2AgCl(s) + Cd(s) \rightarrow 2Ag(s) + 2Cl^- + Cd^{2+}$$

$\Delta^\circ G = -150\ \text{kJ mol}^{-1}$ (the change in Gibbs energy per mole of reaction)

Exergonic means the reaction is spontaneous as written while **endergonic** means that it is not spontaneous; the terms are analogous and related to exothermic and endothermic, respectively, for changes in enthalpy during reactions.

The reaction as written is **exergonic** (i.e. $\Delta^\circ G$ is negative) under standard conditions; that is to say that the reaction will proceed spontaneously as written.

In a Galvanic cell (Fig. 2.3), the:

i) (net) oxidation occurs at the anode

ii) (net) reduction occurs at the cathode

The letter n is used here to denote the number of electrons transferred in the balanced redox equation. It is a dimensionless number and arises from the fact that each elementary step involves single electron transfers, since electrons are not transferred as pairs. Hence the letter n, which can also be written as ν, represents the number of elementary electron transfer events in the reaction overall. We shall return to this number later in Chapter 3.

AgCl is an insoluble solid on the silver electrode and cadmium is a solid metalloid. Since they are both solids, direct physical contact can be prevented by holding them apart and hence a reaction between AgCl and Cd is also prevented. The electrons must be transported around the circuit rather than transferring directly from the solid Cd to the solid AgCl. The chemical reaction is spontaneous

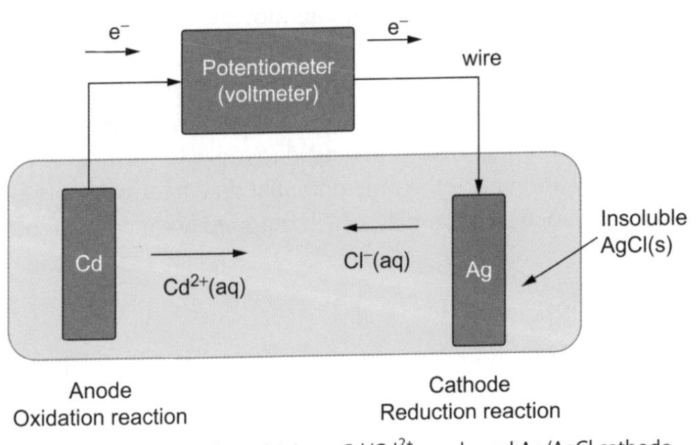

Figure 2.3 Galvanic cell combining a Cd/Cd^{2+} anode and $Ag/AgCl$ cathode.

(negative $\Delta^{o}G$) and electrons are pushed around the external circuit producing an EMF (electromotive force), also referred to as a cell voltage.

The voltage is related to the driving force through the relation:

$$\Delta G = -n\,F\,E_{cell}$$

When rearranged this becomes:

$$E = -\Delta G/nF = -(-150\ \text{kJ mol}^{-1})/(2 * 9.649*10^{4}\ \text{C mol}^{-1}) = +0.777\ \text{J C}^{-1} = +0.777\ \text{V}$$

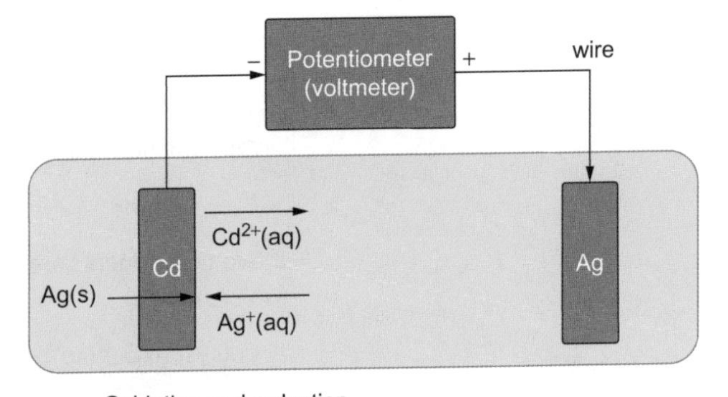

Oxidation and reduction
Reactions occur at the one electrode
So electrons do not need to flow through the wire

The cell potential **does not depend** on the number of electrons transferred in the overall reaction. This is because when we measure the potential of a cell, current is not actually flowing (no net chemistry occurs). Another way to think about it is that the number of electrons transferred per molecule or atom/ion on one electrode is not relevant to the number of electrons per molecule, and so on, transferred on the other electrode.

Whereas with the cell with $Ag/AgCl$ and Cd/Cd^{2+} all components remain separate because AgCl is insoluble, this is not always the case. For example, if $AgNO_3$ (which is soluble) were used instead of AgCl then much less current will flow through the circuit as both half reactions will occur at the cadmium electrode also (i.e. electroless deposition).

The direct reaction between Ag^{+} and Cd can be avoided by physically separating the two reactants using a salt bridge. A salt bridge is a U-shaped tube, usually filled with agar. In a salt bridge K^{+} and NO_3^{-} migrate to maintain electroneutrality. Nowadays, a H^{+} permeable film (e.g. Vycor™ or Nafion™) is often used instead of a salt bridge.

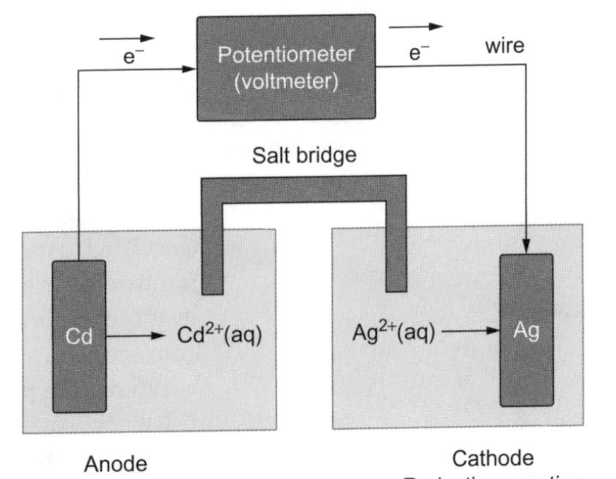

Typically, 3 g of agar agar with 30 g of KCl or KNO_3 in 100 ml of water, which is heated to clarity and cooled in a U-shaped Tube.

Simplistically, the salt bridge works by ions moving into the bridge at one end and 'pushing' ions, already in the bridge, further along until an ion on the other end is pushed out (so the ion that moves in is not the same ion that moves out of the bridge). Since the current that flows is generally quite low (in electrochemical cells), the number of ions that need to move to maintain electroneutrality is quite low and hence the only potential drawback with this approach is the creation of junction potentials, which fortunately cancel each other out.

We will return to junction potentials in Chapter 3.

2.3.1 **Line notation**

A shorthand method for describing electrochemical cells is to use line notation. The convention is to write the anode reaction at the extreme left-hand side and the cathode reaction at the right-hand side. A single line '|' indicates a phase boundary while a double vertical line '‖' indicates a salt bridge or a Nafion film, and so on; for example

$Cd(s)|CdCl_2(aq)|AgCl(s)|\ Ag(s)$ no salt bridge

Anode Cathode

$Cd(s)|Cd(NO_3)_2(aq)\|AgNO_3(aq)|\ Ag(s)$ contains a salt bridge

Anode Cathode

If two components are in the same phase, then they are separated by commas; for example

$Cd(s)|Cd(NO_3)_2(aq)\|Fe^{2+}(aq),\ Fe^{3+}(aq)|\ Pt(s)$

Anode Cathode

Each side of the electrochemical cell is a half-cell and has an associated half-cell reaction.

2.3.2 **Half-cell reactions and standard (reduction) potentials**

This is known as the Gibbs–Stockholm convention.

Half-cell reactions are, by convention, written as reductions:

$AgCl(s) + e^- \longrightarrow Ag(s) + Cl^-$

$Fe^{3+}(aq) + e^- \longrightarrow Fe^{2+}(aq)$

Each of these reactions have an associated half-cell reduction potential (E^o) when measured under standard conditions (293 K, 1 atm, unit activity) or rather calculated to those conditions since very often it is not feasible to measure under standard conditions. For solids and bulk liquids activity is unity. For gases, activity is proportional to the partial pressure and for an analyte, it is proportional to concentration at low concentrations.

What is the potential of a half-cell?

It is not possible to measure a half-cell reaction on its own but we take the convention that the half-cell potential is measured with the other half-cell being the SHE or NHE (standard or normal hydrogen electrode), which by definition has $E^o = 0$ V at all temperatures. The choice of the NHE is not entirely arbitrary as it is an easily constructed half-cell that is highly reproducible. Nowadays, however, it is rarely used since more convenient electrodes such as the saturated calomel electrode (SCE) are available (see Chapter 3).

For example:

$Pt(s)|H_2(g, 1\ atm)|H^+(aq, 1\ M)||Ag(NO_3)(aq, 1\ M)|Ag(s)$

NHE (by default $E^o = 0$) ‖ reaction under examination

The EMF (voltage) of this cell is, by definition, the standard reduction potential for the reaction on the right-hand side (0.799 V).

For an overall reaction the net driving force is the difference between the reduction potentials of each of the half-cell reactions (see Box 2.1). It is essential to realize that the reduction potential is not changed by multiplying the half reactions to balance the equations.

It is tempting to multiply the half-cell potential by 2 and 5, respectively! This should not be done; the potential has nothing to do with the number of electrons transferred in the balanced equation. Remember that potential is measured under conditions where current (electrons) is not allowed to actually flow.

(A) $MnO_4^-(aq) + 8H^+ + 5e^- \rightarrow Mn^{2+}(aq) + 4H_2O \qquad E^o = 1.507\ V$

(B) $Cd^{2+}(aq) + 2e^- \rightarrow Cd(s) \qquad E^o = -0.402\ V$

Balanced equation: $2MnO_4^-(aq) + 16H^+ + 5Cd(s) \rightarrow 5Cd^{2+}(aq) + 2Mn^{2+}(aq) + 8H_2O$

The EMF of the cell (the cell voltage) is given by:

$$EMF_{cell} = E^o_{cell} = E_{RHE} - E_{LHE} = E_{cathode} - E_{anode} = 1.507\ V - (-0.402\ V) = 1.909\ V$$

(RHE = right-hand electrode, LHE = left-hand electrode)

Box 2.1 An analogy between potential difference and water pressure

The EMF of a battery can be measured using a voltmeter. A voltmeter does not draw current, that is, use energy, and hence there is no change in the composition of each half-cell because of the measurement. This is achieved by using a high impedance resistor. For example, when a 1 V battery is connected to a $10^{13}\ \Omega$ resistor the current, calculated using Ohm's law, $I = V/R = 1/10^{13}$ A or $10^{-13}\ C\ s^{-1}$, is 10^{-18} moles of electrons per second.

Understanding the potential difference generated by two half-cell reactions is sometimes difficult. An analogy can be made with water pressure (again this is a loose analogy and should not be taken to literally!). Let's consider a pipe on either end of which we place water pumps. The pumps always try to push water into the pipe. In the middle of the pipe we put a flexible rubber sheet that prevents the flow of water. The sheet is said to 'impede' the flow of water and is analogous to a resistor with a high impedance. If the pumps are identical then when they are turned on the pressure in each direction will be identical and the rubber sheet will not bend. However, if we swap the pump on the left-hand side for a weaker pump then the pressure from the left will be less than that from the right. Of course, the rubber sheet blocks the flow of water but it can stretch. In this situation, the sheet will be stretched in the direction of the left pump until the elastic force (potential energy) built up together with the pressure from the pump on the left matches exactly the pressure generated by the pump on the right. If we swap the pumps then the opposite will happen. In this example, the water does not flow through the pipe but we can determine the difference in pump 'strength' by measuring how far and in which direction the rubber sheet is distorted and from this we can decide what would happen if we removed the sheet. Similarly, measuring potential (voltage) allows us to determine what could happen if we start drawing current without actually having to draw any current.

The reduction potential is used in each case and the sign (+/–) should not be changed.

Since the E^{o}_{cell} = 1.909 V, the $\Delta^{o}G$ for the reaction is negative and hence the reaction is spontaneous as written.

2.3.3 Number lines

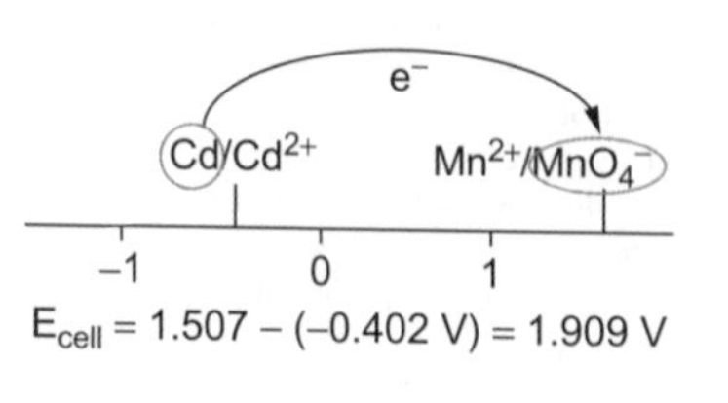

A useful approach to working out which half-cell reaction is the cathode and which is the anode, and whether a reaction will proceed or not for a particular situation is to make use of a simple number line. The number line is constructed as normal with positive potential to the right and negative to the left. The key to remember is that electrons flow from negative to positive.

2.4 Nernst equation

If we know the standard reduction potential for a reaction, what will the reduction potential be under non-standard conditions? The voltage of a battery for example drops at the end of its life; why? When on the top of a ski-piste, the battery in your camera can seem drained while it is not drained at the foot of the ski-piste; why? To answer these questions, we make use of the relation developed by Walther Nernst: the Nernst equation.

Up until now we have been considering reactions under standard conditions (unit activity, 25°C, 1 atm, etc.), however, we rarely if ever use standard conditions (indeed it is often difficult experimentally to achieve such high concentrations).

The Nernst equation (which arises from the mass action law; see the Appendices), allows us to relate the potential of a cell or half-cell to the actual concentration of the species present and the standard reduction potential.

$$aA + ne^- \rightleftharpoons bB \; E^{o} \text{ (half-cell reduction potential)}$$

$$E = E^{o} - \frac{RT}{nF} \ln Q \text{ or } E = E^{o} - \frac{RT}{nF} \ln\left(\frac{\alpha_B^b}{\alpha_A^b}\right)$$

where 'α' is activity in this equation and is related to concentration ($\alpha = \gamma[X]$, with γ the activity coefficient, see Appendix for further details). The term after the natural logarithm is the reaction quotient (Q), which has the same form as the equilibrium constant (although it is not the same: K_{eq} is a specific case of Q). The activity of the reduced species is the numerator and the oxidized species the denominator. Pure solids, liquids, and solvents are not included in the equation as their activity is 1. In practice, we generally do not use the activity of species in solution since at low concentrations the concentration approximates the activity; that is, $\gamma = 1$ in the equation $\alpha_{M+} = \gamma[M^+]$. The concentration of solutes expressed in mol L^{-1} and of gases in bars approximates their activity at low concentrations.

Consider specific situations:

i) A trivial but nonetheless important situation is where $Q = 1$, that is the product of the concentrations of species to the right is equal to the product of all species to the left of the redox reaction, and hence $\ln Q = \ln 1 = 0$. In this situation $E = E^o$.

ii) When the cell potential is zero, that is $E = 0$ V, the cell is at equilibrium and hence the Nernst equation becomes: $0 = E^o - \frac{RT}{nF} \ln Q$ and hence:

$E^o = \frac{RT}{nF} \ln Q$ In this situation, the reaction quotient Q is then the equilibrium constant (K_{eq}).

Example: Write the Nernst equation for following half reaction:

$$\tfrac{1}{4}P_4 = (S, \text{white}) + 3H^+ + 3e^- \rightleftharpoons PH_3(g) \rightarrow Eo = -0.046 \text{ V}$$

The potential of the half-cell (at 298.15 K) can be written:

$E = -0.046 - \frac{0.05916}{3} \ln\left(\frac{P_{PH_3}}{[H^+]^3}\right)$. Note that the concentration of solids is 1 and the concentration of a gas is its partial pressure in bar.

Le Chatelier's principle states that a system at equilibrium when disturbed will adjust its composition spontaneously to reach a new equilibrium position.

To write down the Nernst equation for a complete reaction:

i. Write down the reduction equations and potentials.
ii. Balance the equations and decide which reaction represents the anode and which the cathode (e.g. use a number line).
iii. Use the equation $E^o = E_{cathode} - E_{anode}$.
iv. Decide if the complete cell reaction was written so that the reaction is spontaneous (i.e. ΔG is negative). N.B. that electrons flow towards more positive potentials.

e^- →

Cd/Cd^{2+} Ag/Ag^+

–1 0 1

$E_{cell} = 0.78 \text{ V} - (-0.468 \text{ V}) = 1.24 \text{ V}$

Some examples of the use of the Nernst equation are given in Boxes 2.2–2.4.

Box 2.2 Applying the Nernst equation 1: E^o and the equilibrium constant

When an electrochemical cell is at equilibrium then the current will be zero, as will the cell EMF; that is, a 'dead' battery. Since at equilibrium the driving force for the reaction is zero, $\Delta G = 0$ and hence $-nFE_{cell} = 0$.

For example, if we connect the two terminals of a battery to either end of a resistor (also referred to as a load) it will do work to move electrons through the resistor and the resistor becomes hot due to energy loss. The concentrations of the various components within the battery will then change and eventually the battery will drain: but when exactly? At the point that the battery is drained; that is, at equilibrium;

$E_{cell} = E^o - \frac{RT}{nF} \ln Q = 0$, and therefore

$$E^o = \frac{RT}{nF} \ln Q = \frac{RT}{nF} \ln K_{eq}$$

So, if we calculate the equilibrium constant from the standard reduction potential then we can determine the concentrations of species present at the equilibrium point.

Box 2.3 Applying the Nernst equation 2: E^o and solubility product (K_{sp}):

The reduction potentials for half-cell reactions can be used to calculate equilibrium constants for reactions that are not actually redox reactions themselves. As an example, consider the dissolution of $Fe(CO_3)$ and its solubility product.

$$Fe(CO_3)(s) \rightleftharpoons Fe^{2+} + CO_3^{2-}(aq)$$
$$K_{eq} = K_{sp} = [Fe^{2+}][CO_3^{2-}]/[Fe(CO_3)] = [Fe^{2+}][CO_3^{2-}]$$

This reaction is the sum of two redox reactions, the potentials of which are known:

Reaction 1: $Fe(CO_3)(s) + 2e^- \rightarrow Fe(s) + CO_3^{2-}(aq)$ $E^o = -0.756$ V

Reaction 2: $Fe^{2+}(aq) + 2e^- \rightarrow Fe(s)$ $E^o = -0.44$ V

Reaction 1 - Reaction 2 = $Fe(CO_3)(s) + 2e^- - \{Fe^{2+}(aq) + 2e^-\} \rightarrow Fe(s) + CO_3^{2-}(aq) - \{Fe(s)\}$

And hence $\{Fe(CO_3)(s) \rightleftharpoons Fe^{2+} + CO_3^{2-}(aq)\}$

$= (-0.756\text{ V}) - (-0.44\text{ V}) = -0.31_6\text{ V}$

Since $E^o_{cell} = \frac{RT}{nF}\ln K_{sp}$ and hence $K_{sp} = exp^{\frac{nFE^o_{cell}}{RT}}$

$$K_{sp} = exp^{\left(\frac{(2)(96485\ C\ mol^{-1})(-0.31\ V)}{(8.314\ J\ mol^{-1}\ K^{-1})(298\ K)}\right)} = 3.626 \times 10^{-11} \text{ at } 25^oC$$

This serves to show that electrochemistry is not just about electrons; in many other areas it has real potential as well.

Box 2.4 Applying the Nernst equation 3: E^o and concentration cells

Although it may not seem obvious, it is not necessary to have two distinct reactions in order to create an EMF. In fact, consideration of the Nernst equation can also allow us to determine the difference in chemical potential due only to differences in concentrations of the same species in two different compartments of a cell. Take the following cell for example:

$$Pt(s)|Fe^{2+}(1.0\text{ M, aq}), Fe^{3+}(0.1\text{ M, aq})||Fe^{2+}(0.1\text{ M, aq}), Fe^{3+}(1.0\text{ M, aq})|\ Pt(s)$$

You might expect that the cell potential would be zero as the same reactions occur at both electrodes, however, the system is not at equilibrium since the composition of each cell is not the same. There is a driving force for equalization of the cell compositions. We can show that this is the case using the Nernst equation by determining the potential of each half-cell and then calculating the potential of the cell overall.

$$E_{cell} = E^o - \frac{RT}{F}\ln Q$$

where $E^o = 0.771$ and $n = 1$

$$E_{LHE} = 0.771 - \frac{RT}{F}\ln\frac{1.0}{0.1}$$

$$E_{RHE} = 0.771 - \frac{RT}{F}\ln\frac{0.1}{1.0}$$

$$EMF_{cell} = E_{RHE} - E_{LHE} = \left\{0.771 - \frac{RT}{F}\ln\frac{0.1}{1.0}\right\} - \left\{0.771 - \frac{RT}{F}\ln\frac{1.0}{0.1}\right\}$$

$$= 0.771 - \frac{RT}{F}\ln\frac{0.1}{1.0} - 0.771 + \frac{RT}{F}\ln\frac{1.0}{0.1}$$

$$= -\frac{RT}{F}\ln\frac{0.1}{1.0} + \frac{RT}{F}\ln\frac{1.0}{0.1}$$

$$= \frac{RT}{F}\ln\frac{1.0}{0.1} + \frac{RT}{F}\ln\frac{1.0}{0.1}$$

$$= 2\frac{RT}{F}\ln\frac{1.0}{0.1} = 2\frac{8.314 Jmol^{-1}lK^{-1} \times 298K}{96485Cmol^{-1}}\ln\frac{1.0}{0.1}$$

$$= 0.118 \text{ V or } 118 \text{ mV}$$

This is a relatively large difference and is easily measured. We shall encounter concentration cells again in the context of junction potentials in Chapter 3.

Remember that $-\log\frac{a}{b} = \log\frac{b}{a}$

2.5 Formal oxidation states and reduction potentials

The oxidation state of a metal is determined by the number of electrons it has lost or gained relative to its atomic state. Most transition metals have several oxidation states that are normally accessible [e.g. Mn(0), Mn(II), Mn(III), Mn(IV), Mn(V), and Mn(VII)]. In terms of electrochemistry, oxidation leads to an increase in oxidation state while reduction leads to a decrease in oxidation state. For main group elements the oxidation state is relatively easily assigned. Fluorine is the most electronegative element and is most stable in the –1 oxidation state, whereas in the 0 oxidation state it is very reactive, and indeed elemental fluorine (F_2) is a powerful oxidant.

For transition metals the calculation of formal oxidation state is a little bit more complicated. Two complexes with the same oxidation state can have different reduction potentials since the ligands share electron density (stabilization) with the metal. As a result, the electron density on the iron centre in the complexes $[Fe(CN)_6]^{4-}$ and $[Fe(H_2O)_6]^{2+}$, which are both in the +2 formal oxidation state, is different. It is easier to oxidize one of the complexes than the other. This shows the limitation in the use of formal oxidation states of metals to predict their redox reactions when the ligands around the metal centres are different.

An important point to remember, however, is that the overall charge on the complex is **not** the formal oxidation state! $[Fe(CN)_6]^{4-}$ and $[Fe(H_2O)_6]^{2+}$ are both in the Fe(II) oxidation state.

A question that one may pose then is, if $[Fe(III)(CN)_6]^{3-}$ and $[Fe(II)(H_2O)_6]^{2+}$ were mixed in solution what would happen?

We are of course ignoring the fact that dissolving $[Fe(III)(CN)_6]^{3-}$ or $[Fe(II)(CN)_6]^{4-}$ in 1 M aqueous acid would not be such a clever thing to do as it could result in the release of HCN gas. It is better to use the Nernst equation rather than do the experiment to satisfy our curiosity in this case!

Equation 1: $[Fe(III)(CN)_6]^{3-} + e^- \rightarrow [Fe(II)(CN)_6]^{4-}$ $\quad E_{½} = 0.436$ V

Equation 2: $[Fe(III)(H_2O)_6]^{3+} + e^- \rightarrow [Fe(II)(H_2O)_6]^{2+}$ $\quad E_{½} = 0.771$ V

Equation 1 - Equation 2 →

$$[Fe(III)(CN)_6]^{3-} + [Fe(II)(H_2O)_6]^{2+} \rightarrow [Fe(II)(CN)_6]^{3-} + [Fe(III)(H_2O)_6]^{3+}$$

$$E_{cell} = 0.436\ V - 0.771\ V = -0.335\ V$$

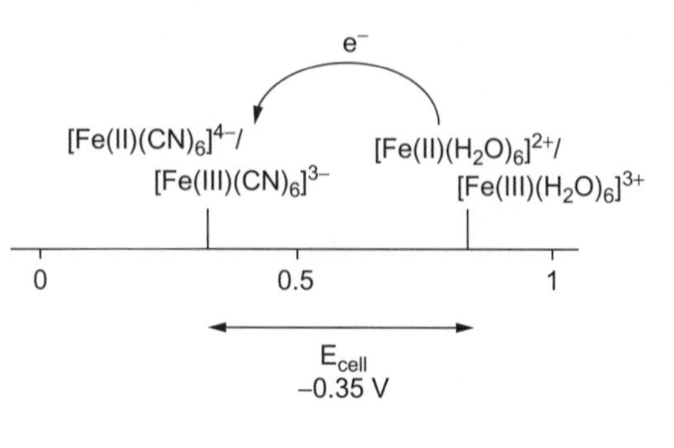

Using the number line approach, we see also that there is a large driving force as the redox potentials are separated by $|E_{cell}|$ = 0.335 V but in this case we have to move an electron towards negative potentials.

Since E_{cell} is negative, ΔG is positive; so when $[Fe(III)(CN)_6]^{3-}$ and $[Fe(II)(H_2O)_6]^{2+}$ are mixed in solution electron transfer is not spontaneous and the oxidation state of either complex will not change. If we mix $[Fe(II)(CN)_6]^{3-}$ and $[Fe(III)(H_2O)_6]^{2+}$, however, there will be a large driving force and the reaction will proceed to, essentially, completion.

2.6 Representing redox chemistry

It is often convenient to use short hand notation and diagrams to represent a large amount of electrochemical data, especially where we can use those representations to predict rapidly the outcome of a particular reaction or situation. In this section, we will focus on four approaches to representing electrochemical data in terms of thermodynamic properties and in particular Gibbs energy (*G*). Latimer and Frost diagrams are relatively simple to construct, requiring a few steps to be followed rigorously in their construction and provide a rapid visual method for determining the stability of various species.

Pourbaix plots are constructed primarily to understand the stability of various species as a function of pH and are useful in allowing us to select appropriate pH ranges in which to work. For example, we can readily see at what pH Fe^{3+} and Fe^{2+} do not exist as individual ions, but instead are oxides (e.g. rust). Finally, we will briefly discuss the construction and use of Ellingham diagrams that allow us to visualize, using electrochemical data, how the spontaneity of a reaction changes with temperature. This diagram is especially useful when recovering metals from metal containing ores and also helps emphasize that we are able to determine not only free energy changes for reactions but also enthalpy and entropy values from electrochemical data.

2.6.1 **Latimer diagrams**

?

+1.589 | +1.154 +1.430 +0.535

$IO_4^- \rightarrow IO_3^- \rightarrow HOI \rightarrow I_4(s) \rightarrow I^-$

+7 +5 +1 0 −1

Formal oxidation state

Latimer diagrams are used to display reduction potentials under specific conditions in a relatively concise manner. An example is the redox chemistry of iodine. The diagram is shorthand for the individual reactions with H^+ and H_2O omitted for simplicity. For example, $IO_3^- \rightarrow HOI$ is shorthand for the half-cell reaction:

$$IO_3^- + 5H^+ + 4e^- \rightarrow HOI + 2H_2O \qquad E^o = +1.154\ V$$

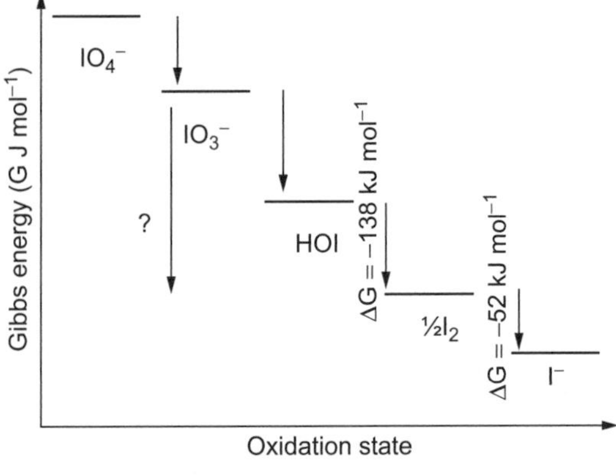

ΔG (?) = −583 kJ mol^{-1}
Hence $IO_3^- \rightarrow ½I_2$ $E_{½}$ = 1.21 V

Reduction potentials that are not shown, that is the one with the '?', can be derived by considering the relation between ΔG and potential ($\Delta G = -nFE$)

The reaction indicated with the question mark is, in full:

$$IO_3^- + 6H^+ + 5e^- \rightarrow I_2 + 3H_2O \qquad E^o = ?\,V$$

The Gibbs energy change in the overall reaction indicated by '?' is the sum of the ΔG of the individual steps.

$\Delta G_{overall} = \Delta G_{step1} + \Delta G_{step2}$ and since $\Delta G = -nFE$ for each step:

$$-n_{overall}FE_{overall} = (-n_{step1}FE_{step1}) + (-n_{step2}FE_{step2})$$

Dividing each term by '$-F$' gives

$$n_{overall}E_{overall} = n_{step1}E_{step1} + n_{step2}E_{step2}$$

and hence:

$$E_{overall} = \frac{(n_1E_1 + n_2E_2 + n_3E_3 + \ldots n_NE_N)}{(n_1 + n_2 + n_3 \ldots\ldots n_N)}$$

where N = a positive integer.

For the reaction

$$IO_3^- + 6H^+ + 5e^- \rightarrow I_2 + 3H_2O \qquad E^o = \frac{(4\times 1.154 + 4\times 1.430)}{(4+1)} = 1.209V$$

An additional use for Latimer diagrams is to assess the stability of a particular species with respect to comproportionation or disproportionation. We will come back to this point in our discussion of Frost diagrams.

2.6.2 **Frost diagrams, disproportionation, and comproportionation**

A Frost diagram illustrates the stability of an element in its various redox states relative to its elemental form. The graph is constructed by plotting $N(E)$ against oxidation state, where N is the oxidation state; the elemental form has an oxidation state of 0, and E is the reduction potential.

For the general reduction:

$$X(N) + Ne^- \rightarrow X(0)\ E^o$$

$$N(E^o) \propto -\Delta_r G^o$$

The most stable oxidation state lies lowest in the diagram and by definition the 0 oxidation state has $N(E^o) = 0$ (since $N = 0$)

Take iodine as an example. The Latimer diagram includes the reduction potentials for all reactions directly to the elemental form of iodine (I_2).

+1.318
+1.209
+1.589 +1.154 +1.430 +0.535
$IO_4^- \rightarrow IO_3^- \rightarrow HOI \rightarrow I_4(s) \rightarrow I^-$

+7 +5 +1 0 −1
oxidation state

Latimer diagram for iodine

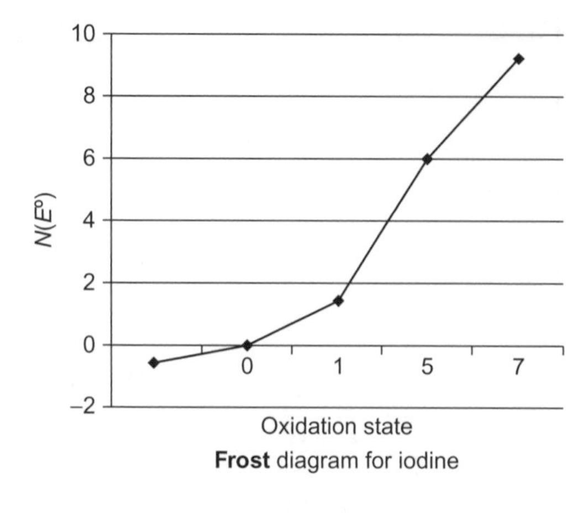

Frost diagram for iodine

The Frost diagram is useful to determine the relative stability of species towards disproportionation and comproportionation.

i) The slope of the line joining two points is equal to the reduction potential of the redox couple between the two oxidation states. The steeper the slope the more positive (or more negative) the reduction potential.

$$\Delta E = \frac{N_1(E^o)_1 - N_2(E^o)_2}{difference\ in\ oxidation\ state}$$

ii) The redox couple with the most positive slope preferentially undergoes reduction and vice versa.

iii) Species undergo disproportionation if the point lies above the line connecting adjacent species and vice versa.

Consider the reaction: $2Cu(I) \rightleftharpoons Cu(0) + Cu(II)$, which is the disproportionation of Cu(I) (the reverse reaction is referred to as comproportionation).

The two half-cell reactions involved are:

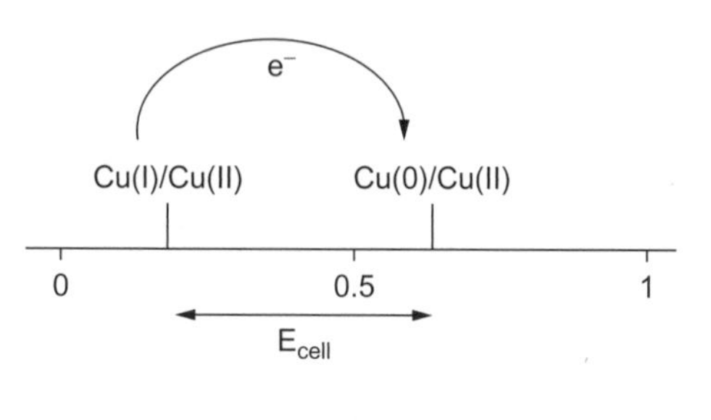

$Cu(I) + ^{-} \Leftrightarrow Cu(0) \rightarrow E^o = 0.521\ V$

$Cu(II) + e^{-} \Leftrightarrow Cu(I) \rightarrow E^o = 0.153\ V$

The number line shows immediately that the overall reaction is spontaneous and:

$E^O_{cell} = E_{cathode} - E_{anode} = 0.521\ V - 0.153\ V = 0.362\ V$

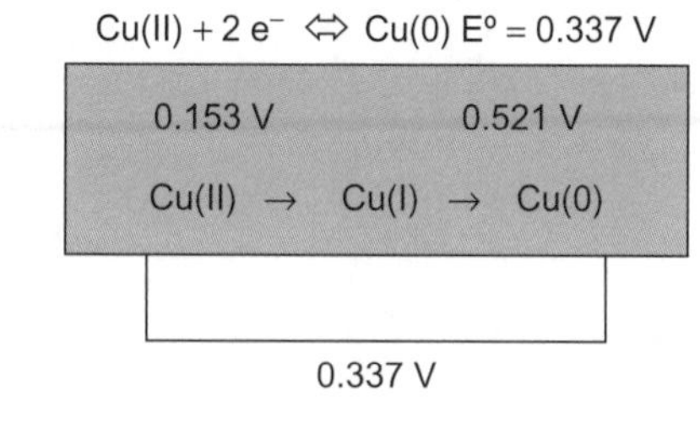

Hence the disproportionation is spontaneous since $\Delta G = -1 * 96\,485$ C $mol^{-1} * 0.368\ V = -35\ kJ\ mol^{-1}$.

The Latimer (left) and Frost (right) diagrams for copper also show thio.

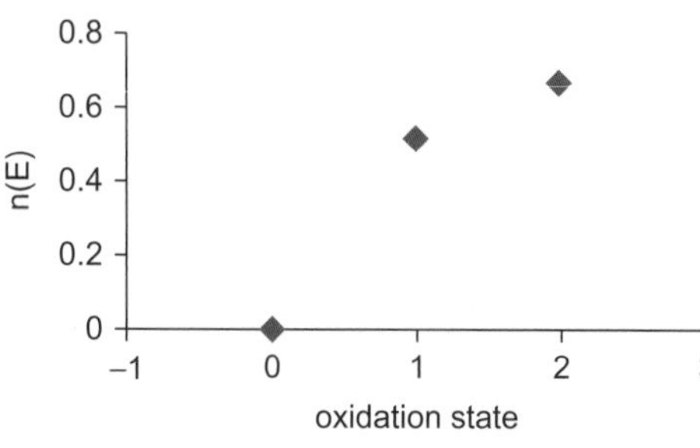

2.6.3 **Pourbaix diagrams and the stability field of water**

A plot of reduction potential vs pH is referred to as a Pourbaix diagram or Pourbaix plot. It is a useful way of representing the pH dependence of reduction potentials graphically and from this the stability of varies species (redox states) as a function of potential and pH. It can also be used to determine at what potential the driving force for a redox reaction (ΔG) becomes negative (spontaneous).

For example, the reduction of Fe^{3+} to $Fe^{2+}(aq)$ does not involve protons:

$Fe^{3+} + 2^{-} \rightarrow Fe^{2+}(aq) \qquad E_{½}^{o} = 0.771\ V$

and hence the reaction proceeds at the same potential regardless of pH.

The reduction of H^+ and oxidation of H_2O to O_2 are the limiting processes in aqueous solution and both reactions are pH dependent.

$H^+(aq) + e^- \rightarrow \tfrac{1}{2}H_2(g)$ $\qquad E_{½}{}^{o} = 0.00\ V$

$O_2(aq) + 4H^+(aq) + 4e^- \rightarrow 2H_2O(aq)$ $\qquad E_{½}{}^{o} = 1.23V$

The Nernst equations for each half reaction are:

i) $$E_{\frac{1}{2}} = E^{o}_{\frac{1}{2}} - \frac{RT}{nF}\, ln\frac{[H_2(g)]^2}{[H^+(aq)]}$$

Since the concentration of $H_2(g)$ does not change (only the proton concentration changes with pH), it has unit activity (1 atm) and $E_{½}{}^{0} = 0$, then the equation simplifies to:

$ln\, X = log_{10} X/log_{10}\, e$ and $-ln(1/x) = -ln(^{-1}) = ln(x)$

$$E_{\frac{1}{2}} = 0 - \frac{RT}{nF}\, ln\frac{1^2}{[H^+(aq)]} = \frac{RT}{nF}\, ln[H^+(aq)] = \frac{RT}{nF}\, log_{10}[H^+(aq)]/(log_{10}\, e)$$

$$= -\left(\frac{RT}{nF(log_{10}\, e)}\right)pH$$

At room temperature, this becomes:

$E_{½} = -(0.059\ V)pH$ or $-59\ mV * pH$

ii) $$E_{1/2} = E^{o}_{1/2} - \left(\frac{RT}{nF}\right)\ln\left(\frac{[H_2O]^2}{[O_{2(g)}][H^+(aq)]^4}\right)$$

Since the concentrations of $O_2(g)$ and H_2O do not change (only the proton concentration changes with pH) and are unit activity (1 atm) then the equation simplifies to:

i) $$E_{\frac{1}{2}} = 1.23V - \left(\frac{RT}{nF}\right)\ln\left(\frac{1}{[H^+(aq)]^4}\right) = 1.23V - \left(\frac{RT}{4F}\right)\ln\left([H^+(aq)]^{-4}\right) = 1.23V -$$
$$(-4)\left(\frac{RT}{4F}\right)\ln([H^+(aq)]) = 1.23V + \left(\frac{RT}{F}\right)\ln([H^+(aq)]) = 1.23V +$$
$$\left(\frac{RT}{F}\right)\frac{\log_{10}([H^+(aq)])}{\log_{10} e} = 1.23\ V + \left(\frac{RT}{F\log_{10} e}\right)\log_{10}([H^+(aq)]) = 1.23\ V - \left(\frac{RT}{F\log_{10} e}\right)pH$$

At room temperature this becomes:

$E_{½} = 1.23\ V.-(0.059\ V)pH$ or $-59\ mV * pH$

Both half-cell reactions vary in exactly the same manner with changes in pH. Indeed, for any reaction where a proton transfer occurs simultaneously with an electron transfer then the reduction potential will vary (by 59 mV) per pH unit at room temperature (Fig. 2.4).

Figure 2.4 shows how the reduction potentials of oxygen and protons vary with pH. The E_{cell} for the reaction $H_2 + ½O_2 \rightarrow H_2O$ does not change with pH, however.

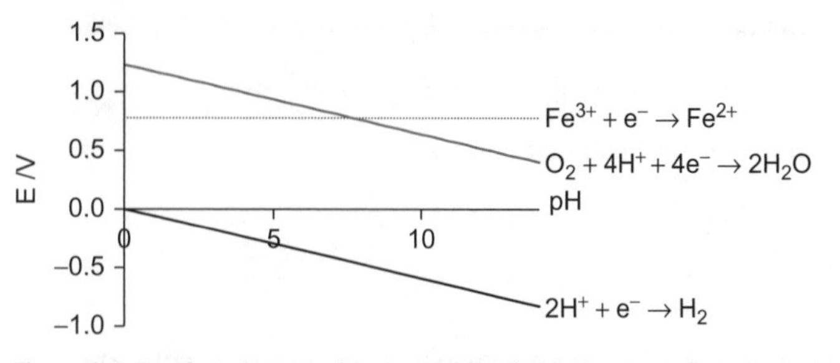

Figure 2.4 Pourbaix diagram showing stability field of water and a pH independent M^{2+}/M^+ redox potential (in reality, the Pourbaix plot for iron is far more complex).

By contrast the reduction potential of Fe(III) is independent of pH. Hence, below.around pH 8, both Fe(II) and Fe(III) are stable in water. What happens above pH 8? Will Fe(II) be oxidized to Fe(III) spontaneously?

To ask the question in another way; at what pH will the oxidation of Fe(II) and reduction of oxygen be at equilibrium (i.e. about to occur spontaneously) under otherwise standard conditions.

The cell reaction would be:

$$4Fe(II) + O_2(g) + 4H^+ \rightarrow 4Fe(III) + 2H_2O$$

The anode reaction is $Fe(III) + e^- \rightarrow Fe(II)$ $\quad E_{½} = 0.771$ V

The cathode reaction is

$O_2(aq) + 4H^+(aq) + 4e^- \rightarrow 2H_2O(aq)$ $\quad E_{½} = 1.23\ V - 0.059\ V \times pH$

$E_{cell} = (1.23\ V - 0.059\ V \times pH) - 0.7771\ V = 0$ that can be rearranged:

$$pH = (0.7771 - 1.23)/(-0.059) = 7.67$$

above this pH, Fe(II) is oxidized spontaneously.

2.6.4 **Ellingham diagrams**

Industrial production of metals from metal salts and oxides is an extremely important component of the modern economy and is also one of the major consumers of energy.

Ellingham diagrams are used to determine the temperatures at which chemical oxidation and reduction reactions will take place. In many cases, especially with steel, coke (carbon) is used as a reducing agent.

$$Fe_2O_3(s) + 3C(s) \rightarrow 2Fe(s) + CO(g)$$

However, mixing iron oxide (rust) with carbon at room temperature will not make steel! The reactions have to be carried out at high temperatures in order for them to proceed. But why?

The reason is that at room temperature the reaction is not spontaneous ($\Delta G > 0$). Remember that the driving force for a reaction must be negative to be spontaneous.

However, ΔG is dependent on temperature: $\Delta G = \Delta H - T\Delta S$ and the reaction shown here has positive entropy; that is, two solids react to give a solid and a gas so disorder (entropy) increases.

As the temperature increases then the negative contribution of entropy to ΔG will increase and eventually overcome the positive enthalpy change (endothermicity, $\Delta H = +ve$) of the reaction. Above this temperature, the reaction will be spontaneous. The graph of $\Delta_r{}^oG$ versus temperature is an Ellingham diagram.

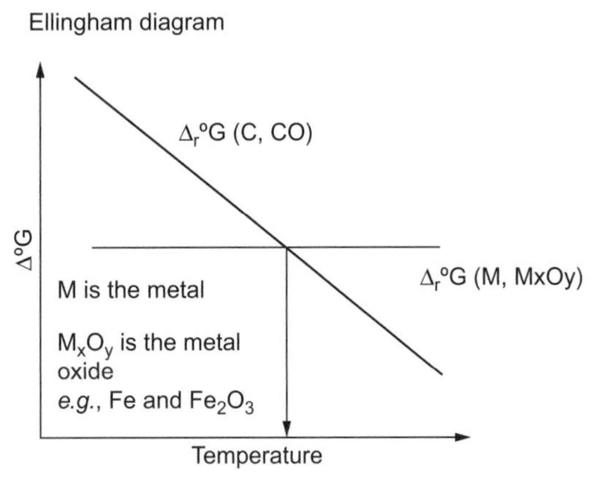

2.7 Summary

This chapter should provide you with an understanding of:

- polarization and depolarization of electrodes and the relation between capacitors and electrochemical cells
- electrodes of the 1st, 2nd, and 3rd kind
- Galvanic (voltaic) cells and calculating cell potentials for combinations of half cells using standard reduction potentials
- the function of salt bridges and electrochemical cell line notation
- the use of the Nernst equation to determine redox potentials under non-standard conditions
- the relation between cell potential and equilibrium constants
- representation of electrochemical properties in Latimer, Pourbaix, Ellingham, and Frost diagram.
- predicting disproportionation and comproportionation reactions

2.8 Exercises

2.1. Explain Faraday's Laws and propose experiments you could do to verify their validity.

2.2. Calculate the mass of nickel deposited on a cathode by electrolysis of nickel nitrate solution using a current of 500 A for 10 h. What is the reaction that takes place at the anode when the nickel is deposited at the cathode?

2.3. Look up the electrochemical series. What are the half reactions involved and the standard potentials for the electrolysis of ia) water and ib) sodium chloride (remember these are each a set of two half-cell reactions)?

2.4. When iron rusts what is the anode reaction and what is the cathode reaction?

2.5. Write down the chemical equations for the following reactions:

a) Oxidation of glucose with $KMnO_4$ (aqueous) in the presence of sulfuric acid to form gaseous CO_2.

b) Oxidation of $FeSO_4$ with $KMnO_4$ in an acidic medium environment (sulfuric acid) to form $Fe_2(SO_4)_3$.

For the second reaction, decide on the anodic and cathodic reactions and draw the cell diagram in line notation. Derive an expression for electromotive force E of this cell.

Gibbs free energy ΔG at constant pressure and temperature is related to the EMF of a reaction. Determine ΔG under standard conditions.

2.6. The Latimer diagram for manganese in acidic media is:

a) Determine the standard reduction potential of the redox couple MnO_4^{2-}/MnO_2

b) Determine the standard reduction potential of the redox couple MnO_2/Mn^{3+}

c) Determine the Gibbs free energy of the following reaction:

$$3MnO_4^{2-} + 4H^+ \Leftrightarrow 2MnO_4^- + MnO_2 + 2H_2O$$

Is the reaction spontaneous as written?

Calculate the equilibrium constant K_c for the reaction.

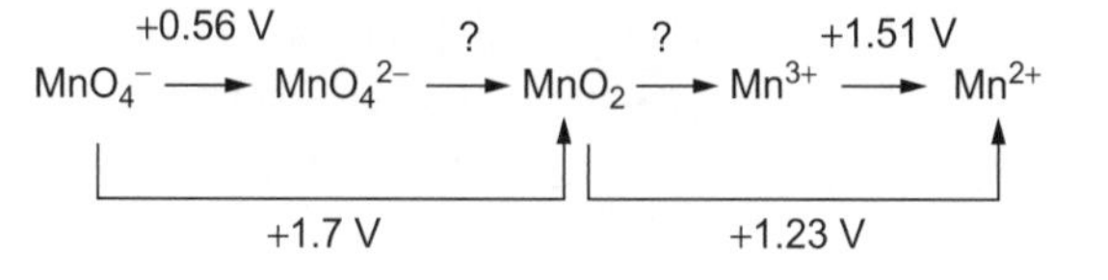

3 Potentials, interfaces, electrodes, and mass transport

3.1 Introduction

Electroanalytical methods can be categorized as static or dynamic. Static implies that the system is not disturbed and we measure the potential of a cell when the current is negligible (potentiometry). We can infer properties from these data; in particular the concentrations of various species. In the dynamic approach, we usually begin with a system at equilibrium; we then disturb the equilibrium by polarizing the working **electrode**, that is, 'apply a voltage' and measure the response of the system (i.e. the charge passed and the time taken to pass it; the current) to achieve the intended change in electrode polarization.

In this chapter, we focus on potentiometric measurements and the electrode solution interface. We begin with the small but extremely important region where the electrode surface meets the solution, the so-called electrode solution interface, containing the double and diffuse layers and the differences between this region and the bulk solution. We then discuss how an electrode and the **double layer** respond to changes in electrode potential. Of course, there are other interfacial regions, such as liquid junctions and membranes, across which potentials can build up and release (which is happening between your synapses hopefully as you read this text!). The potential gradients that are generated when solutions of different compositions are separated by an ion permeable interface will be built on to describe indicator electrodes and in particular the ubiquitous pH meter. We will then discuss how potential can be used in titrations before returning to thermodynamic aspects of electrochemistry. Finally, we discuss mass transport at an electrode to provide a foundation for the electroanalytical experiments discussed in Chapter 5.

3.2 When the electrode meets the solution—the electrical double layer

What does the surface of an electrode look like in detail? This question has occupied the minds of a great many scientists over the last two centuries; the most prominent of whom are Helmholtz (by the 1850s), Stern, Gouy, Chapman, and Grahame, each of whom have lent their names to models to

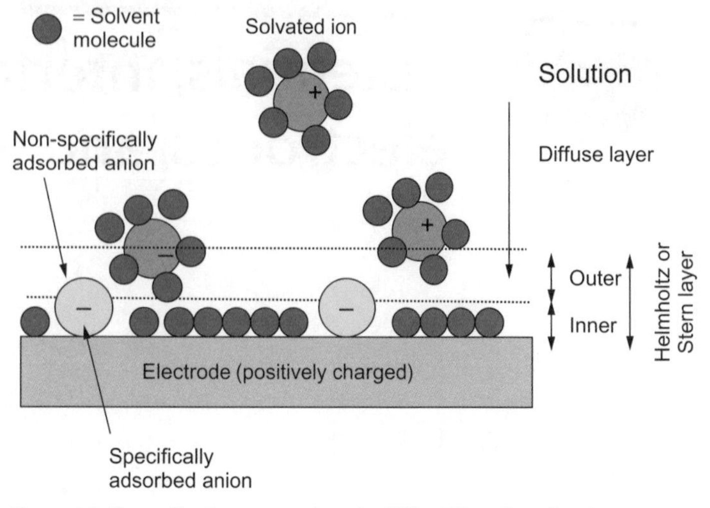

Figure 3.1 Generalized representation of solid liquid interface showing solvent molecules and specifically adsorbed species (in this case anions) on the electrode and solvated ions in solution.

The word **electrode** is used loosely to describe the wire, rod, disc, and so on that we use in an electrochemical experiment, however, it is essential to realize that the 'electrode' is the sum of the solid material and the solution it is in contact with!

describe the electrode solution interface. In this section, we will only focus on the current model.

Consider the surface of gold. It is made up of a layer of gold atoms in contact with the electrolyte (the solution with its solutes). When we polarize the metal, the surface gains excess charge (either positive or negative) at the outer layer of atoms. This charge has an effect on the solution it is in contact with and the various interactions are shown in cartoon form in Fig. 3.1.

The electrode's potential (Galvanic potential, see Chapter 4), is not usually equal to the electrochemical potential (energy) of the bulk solution. Therefore, if the electrode surface is in contact with a solution there will be a sudden change in potential over less than 0.05 nm—which would be a massive electric field gradient (0.5 GV cm^{-1})! However, the potential gradient is in fact less steep because of the effect the polarization of the metal has on the ions and molecules in solution.

In the bulk solution, the solvent molecules are oriented randomly and at ambient temperature they move and tumble rapidly due to **Brownian motion** and convection. The molecules close to and especially in direct contact with the surface, however, will arrange themselves so that their permanent dipole moment cancels (to some extent) the bulk dipole moment of the electrode and in this way minimize their potential energy (Fig. 3.2). As a result, the solvent molecules touching the electrode are much more ordered than in the bulk solution and this layer of solvent is referred to as the **inner Helmholtz plane** (**IHP**). The orienting effect is felt also by the second and third layer of molecules but becomes rapidly less pronounced as the distance from the electrode increases. The orientation of solvent dipoles is, however, weak and insufficient to compensate for the electric field gradient generated by the difference in electrode and solution potentials.

If the metal is polarized so that it has an excess negative charge, then solvated cations will move from the solution to the surface layer of solvent to compensate this

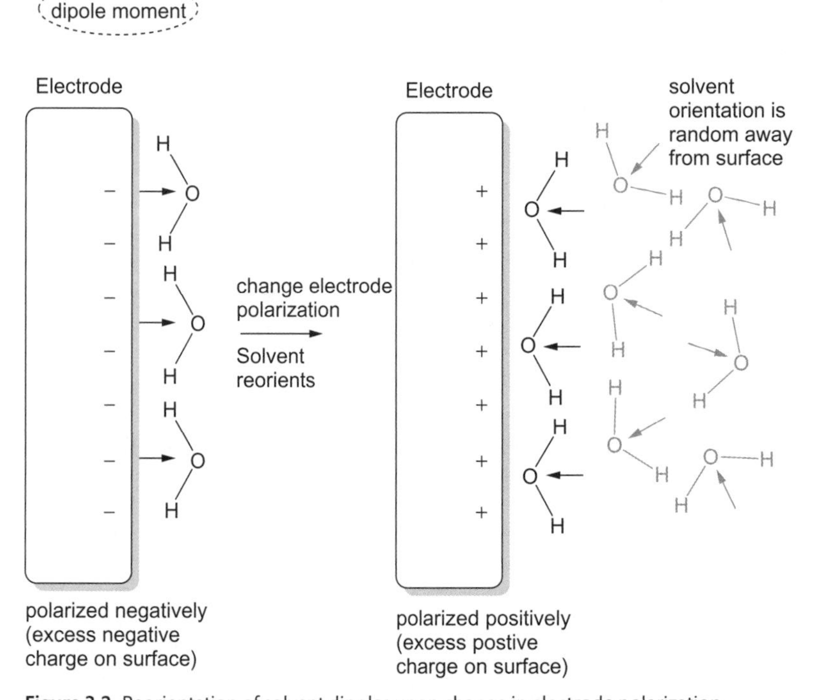

Figure 3.2 Reorientation of solvent dipoles upon change in electrode polarization.

charge and contribute to maintaining local electroneutrality forming what is called the **outer Helmholtz plane** (**OHP**) (Fig. 3.3). The solutes will approach the surface atoms but not touch them because they are themselves surrounded by a layer of solvent molecules (the first solvation sphere). In the Grahame model, some ions are also in direct contact with the electrode (so-called specifically adsorbed ions).

The **solvation shell** is the layer of solvent around a solute. The solute orients such that its dipole moment minimizes the charge density of the dissolved ion or molecule.

The overall arrangement, where the electrode is 'coated' with a layer of oriented solvent, specifically adsorbed ions, and solvated ions at the surface, can be considered as a capacitor with the electrode being one plate and layer of ions the other plate (the Helmholtz model).

The layer of well-ordered solvent molecules and ions is referred to as the **electric double layer**, which is approximately 0.5 nm thick and is formed to render the electrode overall electrically neutral. However, Brownian motion disturbs the order of this layer (the Gouy–Chapman model) and hence the excess charge (e.g. more cations than anions in the case of a negatively polarized electrode) is present further from the electrode forming the **diffuse layer** (not to be confused with the Nernst diffusion layer, Chapter 5). The thickness of the diffuse layer depends on several factors but, under typical electrochemical conditions (0.1 M electrolyte), it is approximately 1–2 nm and the potential changes (or 'drops') over this distance from the electrode.

If the polarization of an electrode is changed then the solvent and ions will respond to counter the change. For example, if the electrode is polarized increasingly negatively, then the solvent will reorient to match their dipole moments to the electrode, positive ions will migrate towards the electrode and negative ions

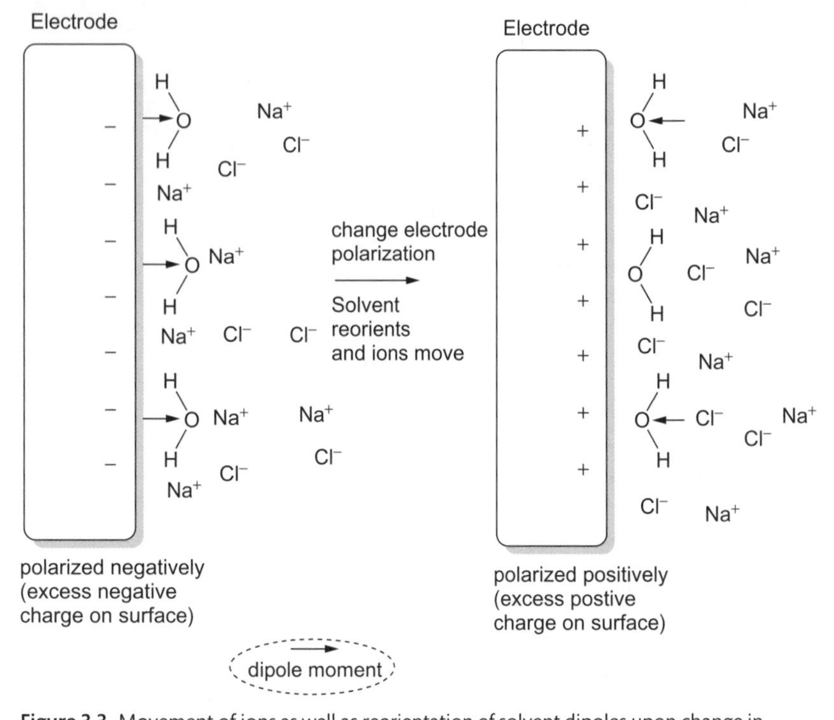

Figure 3.3 Movement of ions as well as reorientation of solvent dipoles upon change in electrode polarization.

away from it. Furthermore, anions will tend to leave the surface (desorb) and cations bind to the surface (adsorb), and vice versa. It should be noted, though, that specifically adsorbed ions can remain adsorbed (Fig. 3.4) even if the electrode potential changes depending on the strength of the interaction between the ion and the surface atoms (see, e.g., Box 5.1).

Overall, even with our simple model of the electrode/solution interface we see that many processes accompany a change in electrode potential. These processes can often be modelled as if they were combinations of simple electronic components (resistors and capacitors), and we will explore this further in Chapter 6.

3.3 Electrode polarization and electrode potentials

In Chapter 2, we introduced the Nernst equation and discussed how the potential of a cell under non-ideal conditions could be calculated if the concentration of each species involved in the overall redox reaction was known. The focus was on electrochemical cells with two electrodes immersed in a conducting solution and connected via an external circuit containing a voltmeter and an ammeter.

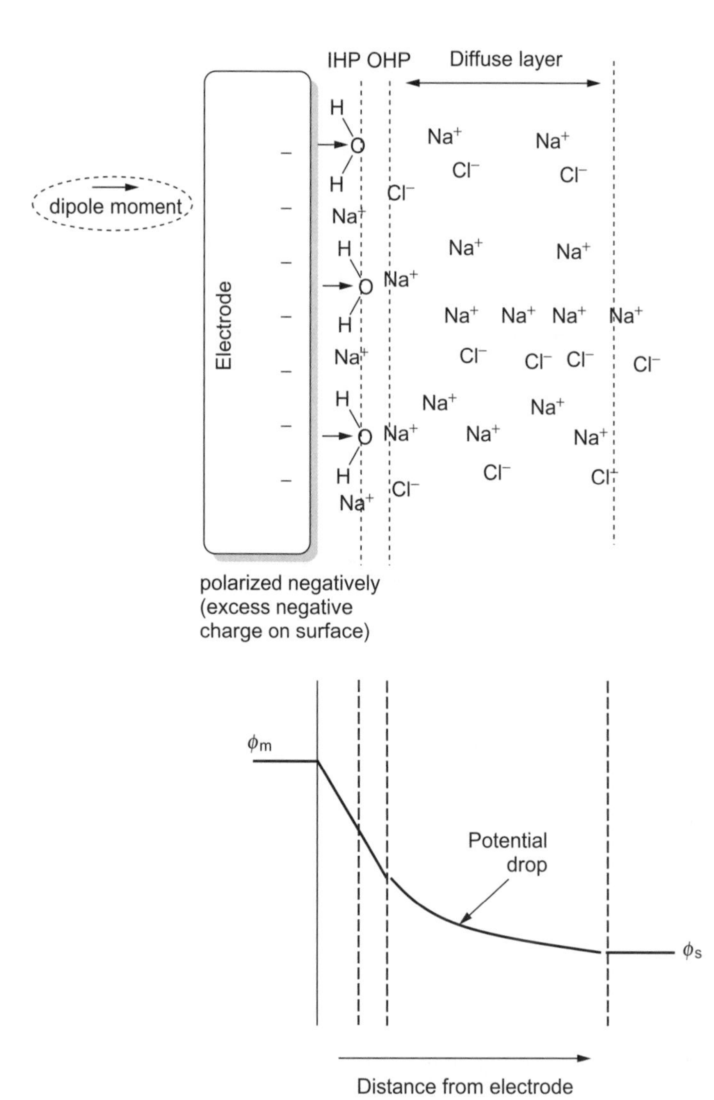

Figure 3.4. Grahame model for the electrode solution interface and the change in potential with distance from the electrode. The potential (ϕ) does not change suddenly at the electrode solution interface but instead the steepest change in potential occurs over the Helmholtz plane with a more gradual drop over the diffuse layer.

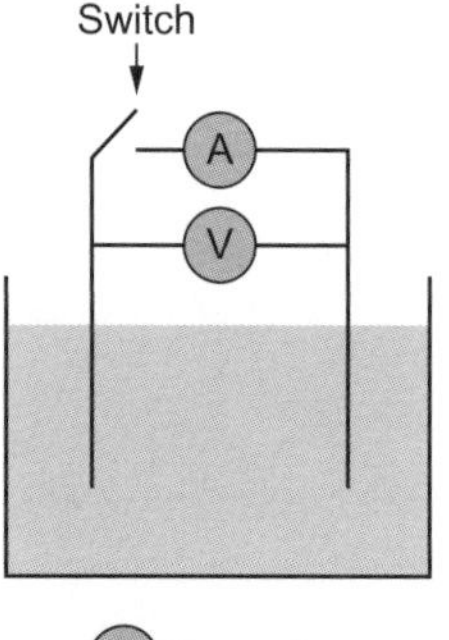

When interested in potentiometric measurements, the switch in the circuit section containing the ammeter is left open. Since a voltmeter has an inbuilt resister with a very high impedance (resistance) and $V/R = I$, the current that can flow will be negligible and the concentration of the various species at the electrodes will not change significantly.

Let us now consider immersing two identical electrodes into the same electrolyte. If the electrodes are connected to a voltmeter then it will show a value of 0.0 V - the difference in the electrochemical potentials of the electrodes is zero ($\bar{\mu}_{LHS} - \bar{\mu}_{RHS} = 0$, see also Chapter 4).

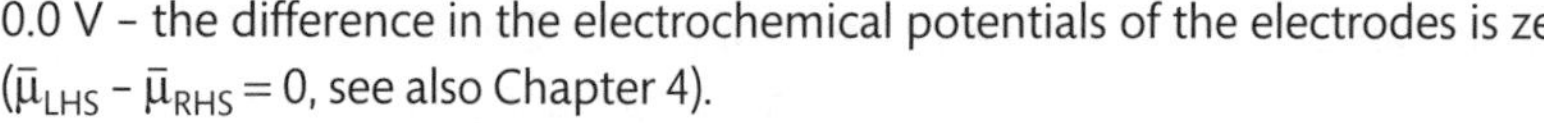

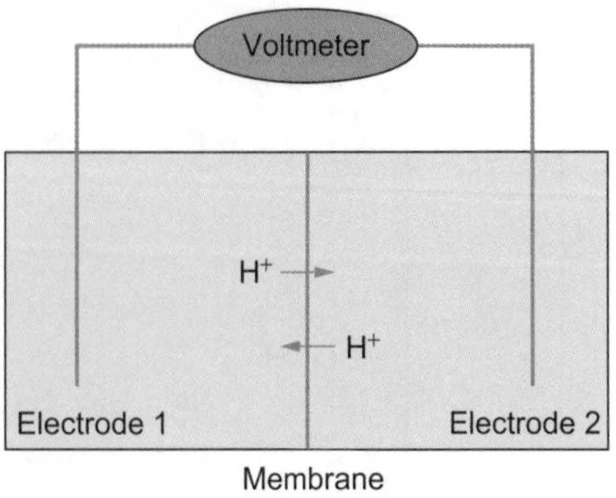

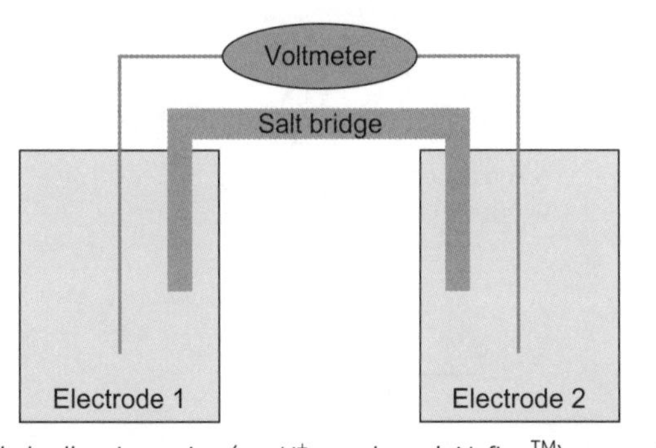

Figure 3.5. Divided cells using an ion (e.g. H^+ pass through Nafion^{TM}) permeable membrane or a salt bridge. The salt bridge is made of a gel that prevents movement of solvent from one cell to the other by convection.

Nafion is a polystyrene, which has been sulfonated heavily, and allows protons but not water or other ions to move through it easily.

Agar agar, also called agar jelly, agar gel, or agar, is made from algae, and when dissolved in water with heating it forms a gel upon cooling trapping any solutes; for example, salts.

In many cases, the components that make up each of the half-cells will react with each other directly. Therefore, the half-cells have to be kept apart physically; either by a semipermeable membrane (e.g. Nafion film, which permits the passage of protons) or a salt bridge (e.g. 3 M KCl in agar agar) as shown in Fig. 3.5 (see also Chapter 2).

3.3.1 Ideal polarizable and non-polarizable electrodes

We polarize an electrode either by removing or adding electrons to it using an external power supply. There are two extremes to the way an electrode responds to our attempts to polarize it–either the electrode changes its potential proportionally to the charge removed (**an ideal polarizable electrode**) or a redox reaction occurs at the electrode to cancel our efforts to polarize it (**an ideal non-polarizable electrode**). We have already met examples of electrodes that come close to each ideal behaviour (a carbon electrode and a Ag/AgCl electrode, respectively).

The glassy carbon electrode in water is an example of an almost ideal polarizable electrode. Carbon is an inert material and polarization to positive or negative potentials can be done with minimal current since there are no chemical

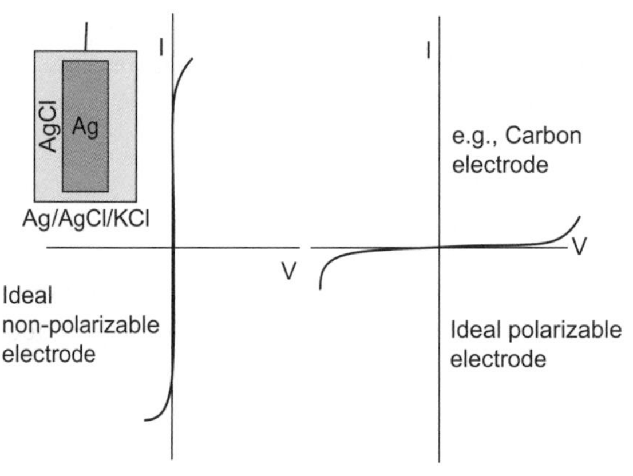

Figure 3.6 *'IV'* curves for an ideal non-polarizable and an ideal polarizable electrode. Note that although we normally treat graphs in which a change made along the horizontal axis drives a change in the vertical axis, in these cases it is the change along the vertical axis (current), which drives a change along the horizontal axis (potential). The lines curve at the end because of a loss of ideal behaviour (see text).

reactions that will 'depolarize' or work against our efforts to draw charge out of or push charge onto the electrode surface. Of course, beyond a certain polarization water will undergo oxidation or protons will be reduced to hydrogen gas and the electrode will cease to show ideal behaviour (these reactions work against our efforts to polarize the electrode further).

A change in polarization will result in a transient current due to the capacitance of the double layer; see Chapter 5.

The Ag/AgCl electrode, an electrode of the second kind (see Chapter 2), behaves very differently and approximates an ideal non-polarizable electrode. Any effort to polarize this electrode will result in either oxidation of Ag(0) to Ag(I)Cl or reduction of Ag(I)Cl to Ag(0) thereby returning the electrode to its original state of polarization (determined by the Ag/AgCl redox couple).

The characteristic response of ideal non-polarizable electrode to any attempt to change its polarization is an instantaneous rise in current. If the current drawn exceeds the rate at which the Ag and AgCl can respond to depolarize the electrode then it will cease to show ideal behaviour and instead the electrode will become polarized (i.e. the electrode potential will change positively or negatively, Fig. 3.6).

3.4 The liquid-liquid interface and membranes

Although it is tempting to think of an 'electrochemical' interface involving only a solid surface, other interfaces are of interest to electrochemistry, especially those involving membranes; including the lipid bilayer of living cells and mitochondria, and membranes separating solutions with different concentrations of electrolytes. The interface between immiscible liquids is also of importance, for example, water and dichloromethane in separation technologies. When separating half-cells with membranes or salt bridges, junction potentials are created, especially when the concentrations of electrolyte on either side of the membrane are different or where salts in which one ion diffuses faster than the other (e.g. LiBr) are used. In the latter case, we need to consider transport numbers also. In this section, we will look briefly at the consequences of differences in the concentrations of electrolytes in pairs of half-cells.

3.4.1 **Transport or transference numbers**

Transport numbers (t_- and t_+) express the fraction of the charge carried by an anion and a cation, respectively—basically the fraction contribution an ion of one type makes to the total conductivity of a solution. The total sum of the transport numbers ($t_+ + t_-$) is 1. For example, the transport numbers of K^+ and Cl^- are approximately equal since both ions are approximately the same size and hence move through solution at equal speeds. In contrast, the transport numbers for H^+ and Cl^- are 0.8 and 0.2, respectively; which is to say that protons move faster than chloride ions and contribute more to the movement of charge (conductivity) in solution.

3.4.2 **Transfer of ions across a phase boundary and liquid junction potentials**

The spontaneous (net) movement of charge across a boundary is driven by a potential gradient. Consider a cell divided by a membrane that permits ions to cross, with pure water on each side and two inert electrodes, one in each chamber. If NaCl is added (to bring the concentration to 0.1 M NaCl) to only one side of the membrane then a sudden change in cell potential will be observed even though a redox active species has not been introduced. The cell we have just made is called a **concentration cell**.

Even if water is 'pure' there are still ions present; that is, 10^{-7} M H^+ and 10^{-7} M OH.

The potential indicated on the voltmeter = $E_{cell} + E_J$ (where E_{cell} is the potential difference between the half-cells in the absence of the membrane). Hence, if the electrodes are identical then the potential, called the Donnan term or Donnan potential, will be determined solely by the activity of the ions (and at low concentrations, $\alpha \approx$ concentration) either side of the membrane.

We have already met concentration cells briefly in Chapter 2 in relation to the Nernst equation. It is useful to remember that the Nernst equation is, first and foremost, a consideration of how the Gibbs energy of a system changes when conditions deviate from non-standard conditions ($\Delta G = \Delta G^o + RT\ ln\ Q$). Hence, if we have a difference in the concentration of electrolyte (even a non-redox active electrolyte such as KCl) on either side of a boundary then we create an unequal situation in terms of Gibbs energy (Fig. 3.7). The relation between Gibbs energy and potential means that electrochemistry is useful for the study of any charged species, both the electrons or ions moving across a phase boundary.

An additional contribution arises from diffusion effects. Different types of ions move at different speeds in solution and through an ion permeable membrane. These differences will have an effect on the potential across the membrane. The speed with which different ions move through a membrane is not a constant

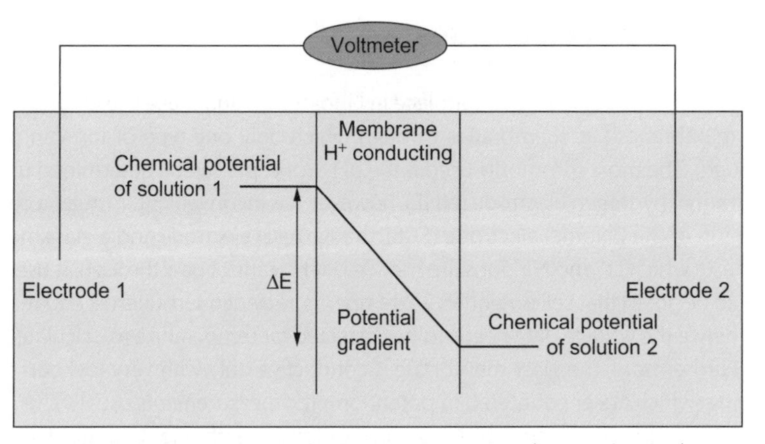

Figure 3.7 If the solutions either side of a conducting membrane (e.g. a Nafion film) are not identical, then their chemical potential (free energy) will be different. This difference is manifest as a potential drop across the membrane (E_j).

but instead is affected by hydration, ion size, the chemical nature of the membrane material, and so on. The diffusion of the smaller Na^+ ions across the membrane will be faster than the Cl^- ions, which will create a local concentration (and charge) difference either side of the membrane. Eventually the charge difference will slow the diffusion of the Na^+ ions to that of the Cl^- ions and the potential will then remain constant.

For example, the transport numbers (or transference number, the fraction of charge carried by each of the ions, see Section 3.4.1) for the cation and anion in KCl and in KNO_3 are approximately equal, and hence the E_j will be close to zero ($t^+ - t^- \approx 0$) and only the contribution of diffusion is important.

In another situation where, for example, $CuSO_4$ is on one side a membrane and $ZnSO_4$ on the other, then the difference in the rate of diffusion of the Cu(II) and Zn(II) ions across the membrane separating the two solutions will generate a transient junction potential also. A junction potential can be minimized by using a salt bridge (see Chapter 2) between cells. The reason this works is that at each end of the salt bridge there will be a large junction potential. However, the two junction potentials will be approximately equal and opposite and hence they will cancel each other out.

The potential difference developed is determined by the difference in charge carried by the cation and anion (transport numbers) and the differences in activity of the left- and right-hand side solutions:

$$E_j = (t^+ - t^-)(RT/F) \ln (\alpha^{RHS}/\alpha^{LHS})$$

In the case where only one ion can move through the membrane, for example, H^+ in the case of Nafion membranes, then its transport number is unity (it is the only ion responsible for carrying charge) and the junction potential is:

$$E_j = (RT/F) \ln (\alpha_{H+}{}^{RHS}/\alpha_{H+}{}^{LHS})$$

For a detailed discussion see Bard and Faulkner, *Electrochemical Methods.*

3.5 Indicator electrodes

The saturated calomel electrode (SCE, $Hg/Hg_2Cl_2/KCl$) is used because it is highly stable and reproducible. The second reference electrode is a Ag/AgCl electrode contained in 1 M HCl, and the potential difference between these two electrodes is by chance close to zero.

Indicator electrodes are commonplace in laboratories and make use of ion selective membranes (i.e. membranes through which only one type of ion can pass through). The most commonly used is the pH probe. pH can be determined using the normal hydrogen electrode (NHE), however, it is inconvenient. Instead a combination of the Calomel electrode (SCE), the Ag/AgCl electrode and a glass membrane, in which Li^+ and Na^+ ions are mobile but H^+ cannot pass through, is used. It should be noted that cell potentials are temperature dependent (Nernst equation!) and hence the voltage data needs to be corrected for temperature to calculate the pH. Furthermore, the glass membrane is conductive only with very low currents flowing (which are encountered in potentiometric measurements usually).

A difficulty with using potentiometric sensors is that high impedance (resistance) terminations are used in the circuitry to minimize the current; however, it is the current that is measured. Hence, the detection circuitry needs to be highly accurate and sensitive over a wide current range (log scale!)

A soft glass (a glass with a high sodium content) membrane separates the Ag/AgCl electrode from the solution being tested. The membrane has a high sodium ion concentration and is conductive to sodium ions. The solution inside the electrode is at a different pH than the solution outside of the membrane and hence a junction potential is created that is related to the pH of the test solution. The technique works because, it is assumed, the surface of the glass has Si-OH units that can deprotonate. Deprotonation creates a junction potential that is measured and can, with calibration, and adjustment for temperature, be used to calculate pH.

3.6 Potentiometric titrations

A common application of potentiometry is to measure the concentration of species in particular oxidation states. For example, if we want to determine the content of Fe^{2+} in a mixture of Fe^{2+} and Fe^{3+} then we can titrate the mixture with a solution of Ce^{4+}, which is a strong oxidant, that is it has a reduction potential much more positive than that of the Fe^{3+}/Fe^{2+} redox couple.

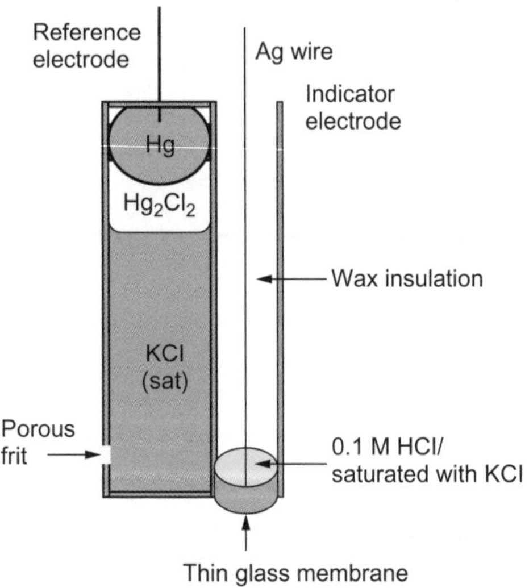

$Hg \mid Hg_2Cl_2 \mid KCl(sat.) \mid \mid H_3O^+ \mid \mid Cl^-(1\ M), AgCl(s) \mid Ag$

$$Fe^{3+} + e^- \rightarrow Fe^{2+} \qquad E^o = 0.77\ V$$

$$Ce^{4+} + e^- \rightarrow Ce^{3+} \qquad E^o = 1.70\ V\ (\text{in } 1\ M\ HClO_4)$$

From the reduction potentials, it can be seen that Fe^{2+} will be oxidized by Ce^{4+} completely (in fact the reaction is not only thermodynamically favourable ($K_{eq} = 10^{17}$), it is also extremely fast) so the overall reaction:

$$Fe^{2+} + Ce^{4+} \rightarrow Fe^{3+} + Ce^{3+}$$

will be essentially complete as soon as the solutions are mixed.

The titration of the Fe^{2+}/Fe^{3+} containing solution with a known Ce^{4+} containing solution (Fig. 3.8) can be monitored by measuring the EMF of the cell shown in Fig. 3.8. The iron ion containing solution being titrated is one half-cell (an electrode of the third kind, $Pt|Fe^{2+}, Fe^{3+}$) and the other half-cell contains a reference electrode (of the second kind, e.g. SCE, Ag/AgCl, etc.). It is essential to remember that, at all times, it is the open circuit potential (the EMF) that is measured, that is, current does not flow through the cell circuit.

It is tempting to think that the redox reaction between iron(II) and cerium(IV) determines the cell potential, but this is incorrect! The question arises then as to what does determine the potential measured.

The E_{cell} is determined by the difference in potential between the reference (e.g. Ag/AgCl) electrode and the Pt/solution electrode. The potential of an electrode of the second kind (i.e. Ag/AgCl) will not change during the experiment. The E_{cell} measured at any time before, during and after the titration, will be dependent on the composition of the titrated solution, which can be calculated using the Nernst equation and the concentrations of any of the pairs of species in solution (e.g. Fe^{3+}/Fe^{2+}, Ce^{4+}/Ce^{3+}). Whether we consider only the ratio of Fe^{3+}/Fe^{2+} or only the ratio of Ce^{4+}/Ce^{3+} depends on convenience and regardless of our choice we will calculate the same potential.

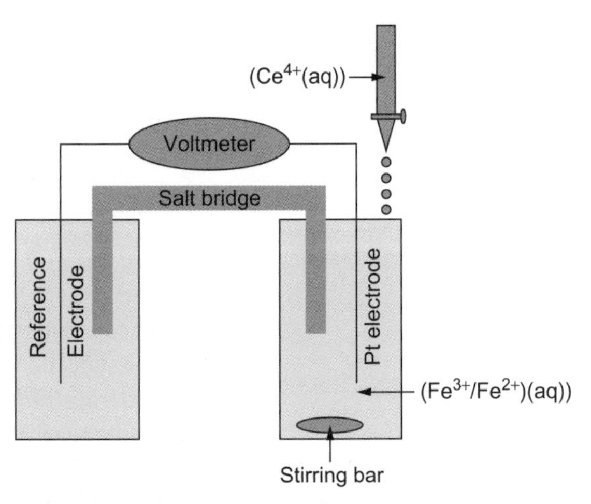

Figure 3.8 Divided cell used in redox titration.

The 'open circuit potential' is the potential measured when the circuit contains a voltmeter only and current cannot pass.

$$E_{cell} = E^o_{\left(Fe^{3+}/Fe^{2+}\right)} - \left(\frac{RT}{nF}\right)\ln\left(\frac{[Fe^{2+}]}{[Fe^{3+}]}\right) \text{ or } E_{cell} = E^o_{\left(Ce^{4+}/Ce^{3+}\right)} - \left(\frac{RT}{nF}\right)\ln\left(\frac{[Ce^{3+}]}{[Ce^{4+}]}\right)$$

Why is this possible? The potential of the electrode of the third kind is solely dependent on the concentrations of the species in solution. The ratio of oxidized and reduced forms of each redox active species at equilibrium is determined by the redox potentials of all species in solution. We can combine the two equations from before to make a single equation that takes all species into account—this may be a bit more intellectually satisfying, but, as we will see, depending on the conditions at a particular point in the titration, we can simplify the combined equation to get back to either of the two previous equations.

$$E_{cell} + E_{cell} = E^o_{\left(Fe^{3+}/Fe^{2+}\right)} - \left(\frac{RT}{nF}\right)\ln\left(\frac{[Fe^{2+}]}{[Fe^{3+}]}\right) + E^o_{\left(Ce^{4+}/Ce^{3+}\right)} - \left(\frac{RT}{nF}\right)\ln\left(\frac{[Ce^{3+}]}{[Ce^{4+}]}\right)$$

And with a trivial algebraic rearrangement:

$$2E_{cell} = E^o_{\left(Fe^{3+}/Fe^{2+}\right)} + E^o_{\left(Ce^{4+}/Ce^{3+}\right)} - \left(\frac{RT}{nF}\right)\ln\left(\frac{[Fe^{2+}]}{[Fe^{3+}]}\right) - \left(\frac{RT}{nF}\right)\ln\left(\frac{[Ce^{3+}]}{[Ce^{4+}]}\right)$$

Inverting the natural logarithmic terms: $\ln\left(\frac{a}{b}\right) = -\ln\left(\frac{b}{a}\right)$

$$2E_{cell} = E^o_{\left(Fe^{3+}/Fe^{2+}\right)} + E^o_{\left(Ce^{4+}/Ce^{3+}\right)} + \left(\frac{RT}{nF}\right)\ln\left(\frac{[Fe^{3+}]}{[Fe^{2+}]}\right) + \left(\frac{RT}{nF}\right)\ln\left(\frac{[Ce^{4+}]}{[Ce^{3+}]}\right)$$

And combining the natural logarithmic terms:

$$2E_{cell} = E^o_{\left(Fe^{3+}/Fe^{2+}\right)} + E^o_{\left(Ce^{4+}/Ce^{3+}\right)} + \left(\frac{RT}{nF}\right)\ln\left(\frac{[Fe^{3+}][Ce^{4+}]}{[Fe^{2+}][Ce^{3+}]}\right)$$

Before we add any Ce^{4+}, the potential will depend only on the concentrations of Fe^{2+} and Fe^{3+} present. When Fe^{2+} is present in solution, in a large excess with respect to Fe^{3+}, then all added Ce^{4+} will react with Fe^{2+} to give Fe^{3+} and Ce^{3+}. The exact concentration of Ce^{4+} remaining will be unknown and well below the limit of quantification. Hence, it is more convenient to calculate the cell potential by considering the Fe^{3+}/Fe^{2+} redox couple. Since the $[Fe^{2+}] > [Fe^{3+}]$ then the potential of the half-cell, calculated using the Nernst equation, will be much less positive than 0.77 V. Each addition of Ce^{4+} containing solution will result in an increase in the concentration of Fe^{3+} at the expense of Fe^{2+}. As the concentration of Fe^{2+} and Fe^{3+} approaches equality then the potential will rise gradually to 0.77 V and stay at close to that value until the concentration of Fe^{2+} falls below 1% of the total iron concentration, after which the potential will rise rapidly to that of the Ce^{4+}/Ce^{3+} redox couple 1.70 V. Since Ce^{4+} is now present in solution at appreciable concentrations, Fe^{2+} is essentially absent.

The reduction potential of Ce^{3+}/Ce^{4+} is more positive than the H_2O/O_2 reduction potential so in principle Ce(IV) should oxidize water. However, although this does happen the rate of the reaction is sufficiently low to allow us to prepare and use solutions in titrations.

At the equivalence point, that is the point at which essentially all Fe^{2+} has been oxidized to Fe^{3+}, we cannot use either redox couple to calculate the cell potential as we cannot know accurately the concentrations of Fe^{2+} and of Ce^{4+} in solution. However, we can still calculate E_{cell} at the equivalence point:

Since $[Fe^{2+}] = [Ce^{4+}]$ and $[Fe^{3+}] = [Ce^{3+}]$ then:

$$\frac{[Fe^{3+}][Ce^{4+}]}{[Fe^{2+}][Ce^{3+}]} = 1 \text{ and } \ln\left(\frac{[Fe^{3+}][Ce^{4+}]}{[Fe^{2+}][Ce^{3+}]}\right) = 0$$

Therefore,

$$2E_{cell} = E^o_{\left(Fe^{3+}/Fe^{2+}\right)} + E^o_{\left(Ce^{4+}/Ce^{3+}\right)} + \left(\frac{RT}{nF}\right)\ln(1)$$

and hence:

$$E_{cell} = \frac{E^o_{\left(Fe^{3+}/Fe^{2+}\right)} + E^o_{\left(Ce^{4+}/Ce^{3+}\right)}}{2} = \frac{0.771\ V - 1.70\ V}{2} = 1.23\ V$$

The potentiometric titration is based on a repeated change in the equilibrium position by addition of Ce^{4+}(aq). After each addition the solution is stirred and after mixing the potential of the half-cell relative to a reference electrode (e.g. Ag/AgCl, SCE, etc.) is measured. The plot of E_{cell} versus volume of Ce^{4+} solution is used to determine the total concentration of Fe^{2+} originally in solution (Fig. 3.9).

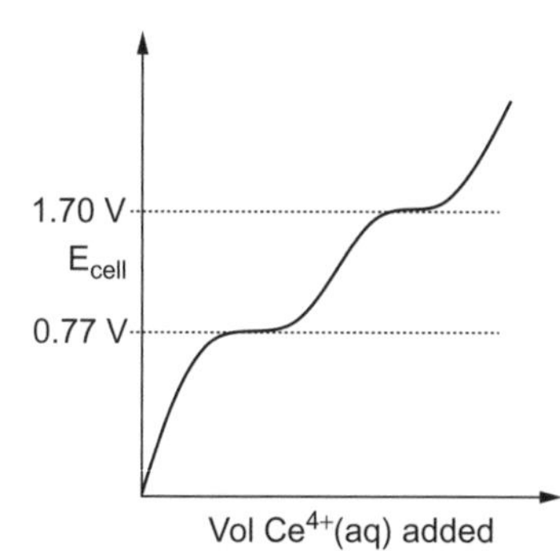

Figure 3.9 Change in cell (Fig. 3.8) potential as Ce^{4+} is added.

3.7 Thermodynamic properties from potentiostatic measurements

Although the recent advent of ab initio calculation methods has enabled thermodynamic data to be predicted relatively easily, there is continued need for experimentally determined thermodynamic data. Calorimetry is the premier method for obtaining thermodynamic data; for example, enthalpy is determined directly by calorimetry as the heat change in an isolated system as a reaction takes place. However, since Gibbs energy changes (ΔG) are directly related to differences in electrochemical potential, electrochemistry is often nearly as useful;

$$\Delta G = \Delta H - T\Delta S = -nFE$$

This relation implies that potential is temperature dependent and that this dependence is due to the entropy of a reaction. If we remember that: $-\left(\frac{\partial G}{\partial T}\right)_p = S$ (one of the Maxwell's relations) then the relation $-\left(\frac{\partial \Delta G}{\partial T}\right)_p = \Delta S$ holds also.

Note that potential is independent of pressure and hence a full derivative replaces the partial derivative.

Therefore, $-\frac{dnFE}{dT} = \Delta S$ and hence $nF\frac{dE}{dT} = \Delta S$ or rearranged: $\frac{dE}{dT} = \frac{\Delta S}{nF}$

Since ΔG can be calculated from electrochemical potentials, the temperature dependence of potential can be used to determine enthalpy and entropy. Hence:

$$\Delta G = \Delta H - T\Delta S \rightarrow nFE = \Delta H - T\Delta S \rightarrow E = -\frac{\Delta H}{nF} + T\frac{\Delta S}{nF} = -\frac{\Delta H}{nF} + T\frac{dE}{dT}$$

Which can be rearranged to:

$$E = \frac{dE}{dT}T - \frac{\Delta H}{nF}$$

that is, the slope of a graph of potential versus temperature is $\Delta S/nF$ and the intercept at 0 K is $-\Delta H/nF$.

3.8 Mass transport at an electrode

Mass transport phenomena play a key role in dynamic electrochemistry and indeed the current that flows in an electrochemical cell is in large part dependent on the rate of transport (movement) of redox active species to and from the electrode. The transport of mass (i.e. ions and redox active species, etc.) to and from the electrode can occur through three mechanisms, all of which play a lesser or greater role in determining the net current under normal circumstances. The 'structure' of the solution at an electrode is of crucial importance to understanding electrochemical data and especially under dynamic conditions, as will be discussed in Chapter 5 and shown schematically in Fig. 3.10.

The time taken for redox active species to move to an electrode is an important source of impedance (called Warburg impedance) and is a source of 'internal resistance' in batteries.

Diffusion—random movement of species in a solution (random walk).

Migration—charged species in an electric field will be move along the electric field gradient due to coulombic forces.

Convection—movement of the bulk solution carries species with it; for example, eddy currents.

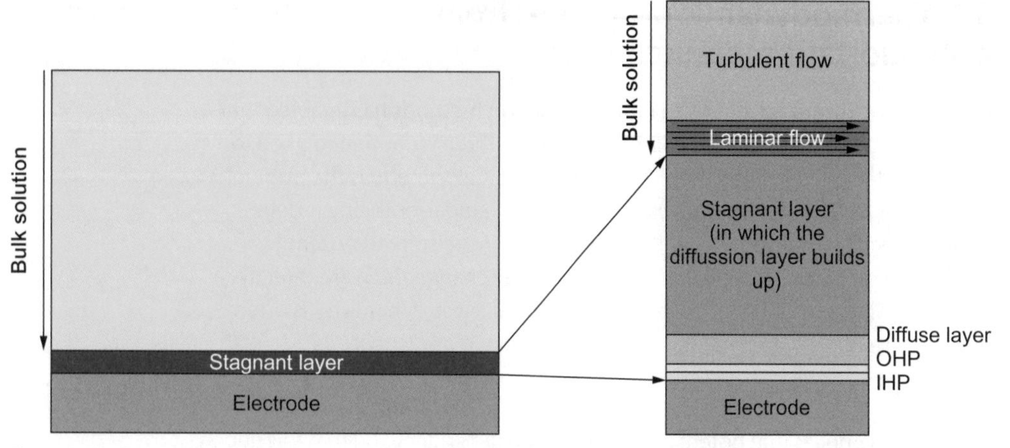

Figure 3.10 The solid liquid interface is comprised of multiple layers. The Helmholtz plane and diffuse region are only up to a few nm thick, however, there is a region that can be up to 50 μm thick depending on solvent viscosity, and so on, called the stagnant layer. This layer is static and mass transport is solely by diffusion. Beyond this layer, mass transport by convection dominates and whole sheets of solvent move in what is called laminar flow, and further still turbulent flow is observed.

3.8.1 **Diffusion**

All ions, molecules, and small particles (e.g. diameter < 500 nm) move continuously (diffuse) in a random walk in solution. Their speed is dependent on their size (hydrodynamic radius), mass, the viscosity of the solution, and temperature. If the concentrations of a species in two regions of a solution are different then entropy will 'force' that the random motion will eventually even out these differences. Electrode reactions occur in proximity to the electrode (i.e. within the **diffuse** layer) and the solution at the electrode is stagnant (does not move). Hence, diffusion is the primary mechanism for mass transport at or near the electrode. If we apply a potential that leads to oxidation of a species at the electrode then we decrease the concentration of the reduced form and establish a concentration gradient (we will discuss the consequences of this in Chapter 5). Diffusion, although random for each individual ion or molecule, will lead to a net movement (flux) of the reduced species to the electrode. Similarly, the oxidized species formed will move away from the electrode because its concentration is highest at the electrode and is zero in the bulk solution.

The thickness of the stagnant layer is ill-defined in reality. For example, the stagnant layer with a solution in a small beaker is essentially the entire contents of the beaker but if there is convection due to evaporation or stirring then the movement of solvent reduces that thickness of the layer and with forced convection it can be as little as a few microns.

The random movement of particles in solution was first noted by Robert Brown (and is now called Brownian motion).

The rate at which species diffuse is described by Fick's 1st and 2nd laws of diffusion. The flux of a species (i.e. the net rate at which the species is moving in solution averaged over a large number of molecules) ***J*** is related to the concentration gradient and the diffusion coefficient ***D***. The first law describes how the flux of a species (J_O) is proportional to the concentration gradient $\left[J = -D\left(\frac{dC}{dx}\right)\right]$.

Fick's 2nd law takes into account that concentration varies with time, that is the rate of change of concentration with time $\left[\frac{dC}{dt} = D\left(\frac{\partial^2 C}{\partial x^2}\right)\right]$ depends on the second derivative of the concentration gradient.

Note that the net movement is towards regions of lower concentration and hence the gradient $\left(\frac{dC}{dx}\right)$ is negative. The minus sign in the equation makes the net flux J a positive number in the direction of lowest concentration.

If we take a large planar electrode, then movement across the electrode can be ignored as the net movement will be zero and only the distance (**x**) from the electrode is important (Fig. 3.11). The rate at which the concentration of a species (**[O]**) changes with time (**t**) depends on the concentration gradient. The steeper the concentration gradient is, the greater the flux will be. Diffusion is usually (but not always) the most important mass transport mechanism for voltammetry and hence, as we will see in Chapter 5, Fick's 2nd law allows us to predict how the concentration will vary with time during an electrochemical experiment.

Any movement parallel to the electrode surface by one molecule is cancelled by the movement of another molecule in the opposite direction, statistically, and hence the only gradient in concentration will be orthogonal to the electrode surface.

3.8.2 **Migration**

Migration is the effect of electrostatic interactions (coulombic forces), which draw positively charged ions to negatively charged electrodes and vice versa. The effect is that the interface between the electrode and the solution becomes charged; ions of one type (either positive or negative) concentrate at the electrode and ions of the opposite charge are repelled (i.e. an electrical double layer is formed). Net movement (flux) due to migration (in one dimension) can be expressed as the contribution to the current by each type of species and described by the equation:

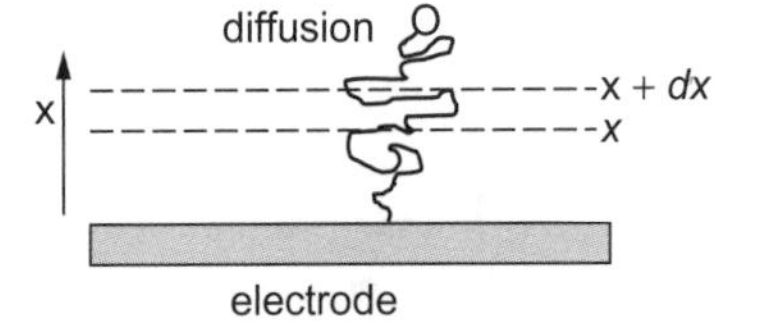

Figure 3.11 Diffusion of molecules is directionally random, but with large numbers of molecules the movements across the plan of a large area electrode cancel out, and only the component orthogonal to the electrode needs to be considered.

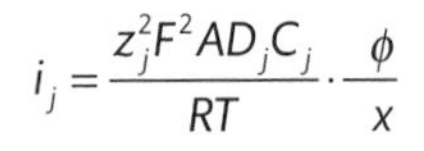

$$i_j = \frac{z_j^2 F^2 A D_j C_j}{RT} \cdot \frac{\phi}{x}$$

where A is the cross-sectional area (cm^2), D is the diffusion coefficient (cm^2 s^{-1}), C is the concentration (M) and $\frac{\partial \phi}{\partial x}$ is the electric field gradient ($\Delta E/l$, V cm^{-1}). Calculating the flux due to migration is complicated when movement by diffusion and convection occurs also. Hence most voltammetric studies are carried out with a redox inert electrolyte (which also reduces cell resistance [impedance]) at high concentrations (0.1–1.0 M). The electrolyte ensures that the strong electric field generated by the application of a potential difference between the electrodes is 'screened' by the electrolyte and hence migration is no longer significant.

Mass transport by migration is not always avoided of course. Indeed, it is the basis of the analytical separation technique electrophoresis, for example, capillary electrophoresis used to analyse mixtures of redox inert ions and gel electrophoresis used to separate proteins or DNA.

3.8.3 **Convection**

Convection is the most important mechanism for mass transport with regard to movement of species over large distances (<0.5 mm). In an apparently static solution, for example, in the absence of stirring, evaporation from the solution's surface cools it resulting in the liquid in contact with the air becoming locally denser and it sinks, establishing a convective current. In general, however, natural (non-stirred) convection is unimportant when experiments take less than 20–30 s in total and hence convection is important only under forced conditions, such as under rapid stirring (as in rotating disc voltammetry) or a jet of liquid (wall-jet voltammetry).

3.8.4 **Laminar and turbulent flow at an electrode**

The movement of solutions at a surface is not the same as in the bulk (Fig. 3.10). In order to understand, for example, the shape of a cyclic voltammogram, it is necessary to consider two types of flow—turbulent and laminar. Laminar (low Reynolds number) and turbulent (high Reynolds number) flow are met more obviously in pipes. If the diameter of the pipe is less than a critical value (dependent on the Reynolds number) then only **laminar flow** will be observed. This dimension depends on the viscosity of the solvent and the flow rate amongst other things. The effect of bulk solvent movement (fluid flow) on the thickness of the stagnant layer can be predictable. For example, if we rotate the electrode rapidly (e.g. 1500 rpm) then the stagnant layer thickness reduces and the diffusion limited current (see Chapter 5) increases. The diffusion limited current varies with the rate of rotation (due to the change in the thickness of the stagnant layer) and forms the basis of rotating disc voltammetry.

In **laminar flow**, solvent moves in a highly ordered way as if the solution was made of thin sheets that glide over each other. In **turbulent flow** the solvent moves in a jumbled manner that provides for more efficient mixing.

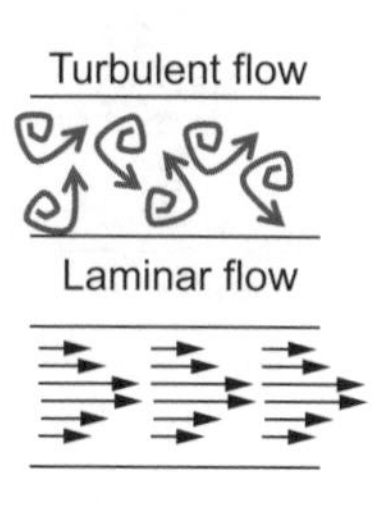

In general, we try to avoid mass transport by more than one mechanism so that only one set of equations is needed to describe a system. So, either we force convection and reduce as much as possible the thickness of the stagnant layer or we keep the setup as still as possible and measure over short time scales with a large amount of supporting electrolyte to exclude convection and migration.

3.9 Summary

This chapter should provide you with an understanding of:

- the structure of the electrode solution interface and the change in potential with distance from the electrode
- electrode polarization and depolarization
- junction potentials
- electrode potentials governed by species at electrodes
- mass transport mechanisms of relevance to electrochemistry (diffusion, convection, and migration)

3.10 Exercises

3.1. Calculate the potential field gradients (V m^{-1}) generated when an electrode is polarized so that its potential is 1 V more positive that the potential of the solution where the thickness of the diffuse layer is 0.01 nm, 0.1 nm, 1.0 nm, and 10 nm. Comment on the effect these differences in gradient would have on ions in solution.

3.2. Consider Fig. 3.5. What value would be indicated on the voltmeter if the solution on the LHS is 0.01 M KCl (aq) and on the RHS is 0.1 M KCl (aq)? Show that the potential increases by each order of magnitude increase in the difference of concentrations between the two half-cells.

3.3. Consider the formation of a self-assembled monolayer of *n*-octanethiol on a gold electrode (the thiol groups form a strong bond to the gold atoms and the alkanes stack up beside each other through van der Waals interactions). What effect would the SAM have on the potential drop at the electrode solution interface and the capacitance of the double layer?

3.4. Predict the potential versus volume plot that would be obtained when $KMnO_4$ is titrated with a reducing sugar such as glucose. Assume that the titration is carried out in water with KNO_3 (0.1 M) at pH 2 with a platinum wire indicator (working) electrode and a Ag/AgCl electrode in a cell divided by a salt bridge.

3.5. Mass transport occurs in solution by any of three mechanisms; diffusion, convection, and migration. (a) Describe, using diagrams, briefly each of the three mechanisms. (b) In cyclic voltammetry (Chapter 5), mass transfer to and from the electrode is ideally under diffusion control. Describe briefly how the effect of convection and migration is minimized experimentally.

Heterogeneous electron transfer and the Tafel equation

4.1 Introduction

In Chapters 2 and 3, the use of electrochemistry (potentiometry) to extract thermodynamic information and in analytical applications, for example, pH determination, and so on, was explored. Electrochemistry can provide kinetic information also. In this chapter, fundamental aspects of electron transfer at the electrode will be consider starting with selected concepts that are key in understanding electrode reactions and electrochemistry in general. The key relation $\Delta G = - \nu FE$ and the Nernst equation that we have assumed to be correct until now will be justified before tackling kinetic aspects of electrode reactions and heterogeneous electron transfer and the symmetry factor α. These concepts are built on in the derivation of two key equations in electrochemistry; the Butler–Volmer and Tafel equations and the Tafel plot. The topics covered in this chapter will be the stepping-off point towards understanding dynamic voltammetry (Chapter 5).

4.2 Electron transfer, concepts, and terms

Electrons move from electrode to solution by tunnelling from the atoms on the electrode surface to a species in solution (reduction) and vice versa (oxidation); a phenomenon called **heterogeneous electron transfer.** The rate of electron transfer depends on the structure of the double layer (Chapter 3), energetics and distance between the electrode and the species in solution. Knowledge of this rate is important because, although we assume that the concentrations of various species at the electrode satisfies the Nernst equilibrium. Furthermore, the electrode and concentrations of species at the electrode take a finite time to readjust (by electron transfer) in response to a change in electrode potential.

'free', or rather solvated, electrons can exist in solution (giving an intense blue colour!), but they are only encountered in extreme situations such as when a solution is exposed to short (ns) intense ionizing pulses of light.

Furthermore, if electrode polarization (potential) is changed rapidly, the rate of electron transfer to and from the electrode may not necessarily increase as expected for the increased driving force. The overall rate (the current) is determined by a combination of factors including the overpotential, concentration, mass transport, changes in surface composition (e.g. during electrodeposition the surface changes), charge trapping ... the list can seem endless and is highly dependent on the situation being investigated. In Chapter 5, the processes that accompany a change in electrode polarization will be discussed, while in this chapter a number of contributing concepts and terms encountered frequently in electrochemistry will be explored.

4.2.1 **Heterogeneous electron transfer**

The simplest model for electron transfer between an electrode and a species in solution is to consider the **Fermi level** of the electrode and the energies of the frontier orbitals of the species being oxidized or reduced. A molecule's orbital energies are essentially 'fixed' and so the equilibrium between the electrode and the species in solution depends on the Fermi level, which can be raised or lowered by polarizing the electrode negatively or positively, respectively, by applying a voltage (V). In Fig. 4.1, the energy of the Fermi level is initially below the energy of the lowest unoccupied orbital but above the level of the highest occupied orbital so electron transfer is not thermodynamically favourable.

In metallic electrodes the outer sphere electrons are delocalized over many atoms and hence instead of having a single energy level, there is a continuum of occupied and unoccupied levels. The **Fermi level** is defined as the highest energy occupied levels state but electrons can enter the metal at any higher energy and leave the electrode from any of the lower energy states.

If the Fermi level is raised above the energy of the lowest unoccupied orbital of the species in solution then an electron can transfer to that orbital. Similarly, if the Fermi level drops below the HOMO level of the species in solution an electron will hop into the electrode. Again, this is a highly oversimplified model for electron transfer. However, although imprecise, it is nevertheless a good jumping off point for discussion. A final point is that electron transfer occurs continuously to and from species in solution (see Fig. 1.5). The **net current density**/current flux (j_o) is the difference between the flux of electrons from the electrode to species in solution (j_c) and the flux to the electrode (j_a), that is $j_o = j_a - j_c$ (see bibliography for reference to a discussion on sign conventions).

Remember that 1 V = 1 J C^{-1}; that is, voltage is the energy required/available to move charge.

The flux is the rate at which electrons move across a boundary and has magnitude both not direction. The difference in direction of anodic (i_a) and cathodic (i_c) currents is taken into account by the minus sign in the equation $j_o = j_a - j_c$.

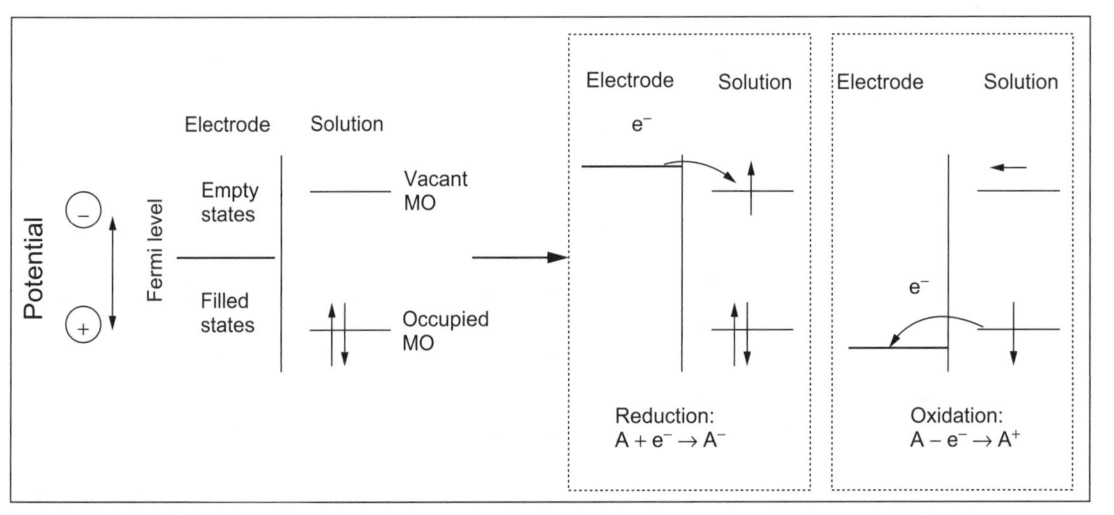

Figure 4.1 Simplified description of changes in the Fermi level of a conductor as an electrode's polarization is changed. The electrons can enter the electrode via any unfilled level and can leave the electrode by an filled level; however, simply think about exchange involving the levels at the Fermi level.

The transfer of electrons from the electrode to species in solution or adsorbed on the electrode surface is 'heterogeneous' since it involves two different phases. The rate constant for electron transfer is '$k^{o\prime}$' if the species involved is in solution and 'k_o' if the species is adsorbed on the electrode.

4.2.2 Equilibrium electrode and formal potentials

The **equilibrium potential** is defined as the potential at which the forward and reverse electron transfer rates are equal and the net current flow is zero. This potential is not necessarily the same as the standard electrode potential (indeed, we seldom if ever measure potentials under standard conditions) and is more precisely referred to as the **equilibrium electrode potential**, which reminds us that the value is specific to the particular electrode that is in use. When the electrode potential is more positive or more negative than the equilibrium electrode potential, the forward and reverse rates are no longer equal and a net current will flow. The current that flows was shown, as early as 1905 by Tafel, to be exponentially dependent on the overpotential provided that the only limit on current flow is the rate of heterogeneous electron transfer. This observation is described by the **Tafel equation**, which we will see can be expressed as $i = a' \exp(b' \eta)$ or $\eta = a + b \log_{10} i$, where η is the overpotential or difference between the applied potential and the equilibrium electrode potential. We will discuss the Tafel equation and its derivation in detail at the end of this chapter.

If we consider a situation where we keep the current sufficiently low and scan rapidly, then we can conveniently disregard mass transport as a limitation to the current measured. We will of course see in the next chapter that we usually cannot do this.

For now, consider the reaction:

$$[Fe(CN)_6]^{3-}{}_{(aq)} + Ag(s) + Cl^- \rightleftharpoons [Fe(CN)_6]^{4-}{}_{(aq)} + AgCl(s)$$

The overall reaction is comprised of two independent reactions that are occurring physically at each electrode with the wires transferring the electrons between the electrodes (see Chapter 2). When the working electrode is polarized (with respect to the reference electrode, e.g. Ag/AgCl), we can calculate the current that will flow to re-establish the relation between E and E^o (i.e. use the Nernst equation)

$$E = E^o_{\left(Fe^{3+}/Fe^{2+}\right)} - \left(\frac{RT}{nF}\right)\ln\left(\frac{\alpha_{Fe^{2+}}}{\alpha_{Fe^{3+}}}\right)$$

where α is activity

At each potential (i.e. E), a different ratio between the oxidized and reduced forms is required in order to be at equilibrium and to reach these concentrations requires the passage of a current (Faradaic) to convert some of the redox active species to a different redox state.

The reaction is at equilibrium when the electrochemical potentials of the reactants and products are equal;

$$\bar{\mu}_{[Fe(CN)_6]^{3-}} + \bar{\mu}_{(e^-)solution} = \bar{\mu}_{[Fe(CN)_6]^{4-}}$$

However, determining E^o is difficult since it requires that all species have an activity of 1 at 25°C and 1 bar pressure. At high concentrations (>0.01 M) activity (α) of ions deviates strongly from concentration and is usually much less than 1.

If H^+ or OH^- is involved in the reaction but not actually being oxidized or reduced, then conditions need to be defined even more precisely. As a consequence, typically we determine the formal potential, which is the potential under a specified set of conditions (e.g. in biology, potentials at pH 7 are more useful than at pH 0, see Chapter 2).

The formal potential (E_f^o) is usually close to, but not equal to, E^o since it is measured under conditions that are non-standard. The difference between E_f^o and E^o is due to differences in the activity coefficients of the oxidized and reduced forms and if protons are involved, the difference in pH is also taken into account.

$$E = E^o - \left(\frac{RT}{nF}\right)\ln\left(\frac{\alpha_{red}}{\alpha_{ox}}\right)$$ where α activity is equal to γC and hence:

$$E = E^o - \left(\frac{RT}{nF}\right)\ln\left(\frac{\gamma_{red}C_{red}}{\gamma_{ox}C_{ox}}\right)$$

by making use of log rules: $E = E^o - \left(\frac{RT}{nF}\right)\left[\ln\left(\frac{\gamma_{red}}{\gamma_{ox}}\right) + \ln\left(\frac{C_{red}}{C_{ox}}\right)\right]$

which when rearranged is: $E = E^o - \left(\frac{RT}{nF}\right)\ln\left(\frac{\gamma_{red}}{\gamma_{ox}}\right) - \left(\frac{RT}{nF}\right)\ln\left(\frac{C_{red}}{C_{ox}}\right)$

and with $E_f^o = E^o - \left(\frac{RT}{nF}\right)\ln\left(\frac{\gamma_{red}}{\gamma_{ox}}\right)$ then: $E = E_f^o - \left(\frac{RT}{nF}\right)\ln\left(\frac{C_{red}}{C_{ox}}\right)$

Overall, for reversible systems, the activity of species in their oxidized and reduced forms tends to be similar and hence the deviation between standard and formal potentials for redox reactions that do not involve protons is usually of the order of a few mV.

4.2.3 **Overpotential**

Although the driving force for electron transfer ($\Delta_r G$) may be negative (and hence spontaneous) it is not necessarily the case that the electron transfer reaction will proceed at an appreciable rate. A so-called activation barrier for the reaction ($\Delta_r G^{\ddagger}$) must also be considered. This barrier can be surmounted by increasing the driving force for the reaction thereby increase the reaction rate, which can be understood using the same approach used in Marcus theory, that is, treating potential energy surfaces as parabolic curves, see Section 4.3. The extra electrode polarization required beyond the formal potential to have a significant reaction rate (i.e. current) is referred to as the overpotential (η) needed; where $\eta = E_{applied} - E_{½}$. It should not be forgotten that, in voltammetry, the potential '0.0 V' means very little except that the polarization of the indicator (working) electrode matches exactly with that of the reference electrode. Of course, we can chose among several reference electrodes that each have their own reduction potentials and quoting a particular potential does not indicate directly the extra driving force available for a reaction. Overpotential is therefore a more useful term since it is directly related to Gibbs energy change and hence allows for facile estimation of the driving force available/needed for a reaction at a particular potential/rate.

4.2.4 The meaning of 'ν'

In the many equations used in electrochemistry the term 'ν' appears and is often replaced by n such as in $\Delta G = -\nu FE$ (often written as $\Delta G = -nFE$ or $\Delta G = -zFE$, depending on the local convention), which is usually taken as the stoichiometry of the overall reaction (number of electrons involved) under consideration. The number is dimensionless (no units such as mole, etc.), and arises because, regardless of the total number of electrons involved overall, each electron is transferred individually from and to the electrode in a series of elementary steps/reactions. Indeed, 'ν' is formally the number of steps leading up to and including the rate determining step in the overall reaction. Furthermore, whereas the potential difference generated by two half-cell reactions is independent of the stoichiometry (e.g. the potential is the same for $H^+ + e^- \rightarrow ½H_2$ as it is for $2H^+ + 2e^- \rightarrow H_2$), the Gibbs energy change (ΔG) depends on the stoichiometry (units of kJ mol^{-1}, referring to per mole of reaction).

With most, but not all, electrochemical reactions, this number is justifiably synonymous with the overall stoichiometry of the reaction in terms of electrons transferred.

4.2.5 Galvani (ϕ), Volta (ψ), and surface (χ) potential

The chemical potential of the solution is dependent on its composition (ionic strength, redox active solutes, etc.) but, for simplicity, we will that assume it is a fixed value during experiments. The chemical potential of the electrode, by contrast, is changed by polarization; positive polarization results in a lowering of its potential, whereas negative polarization increases its potential. Note that we are interested only in the electrical potential at the electrode's surface layer of atoms, which is called its Galvani potential (ϕ_{elec}, or inner electrostatic potential). If we consider each side of the electrode solution interface as described by the Stern model (i.e. the electrode surface and the Helmholtz plane). When an electrode is polarized negatively, that is, the electron density at the surface is high, the solvent and electrolyte is structured such that positively charged ions are attracted to the electrode (IHP/OHP) and the diffuse region with the result that there is an excess positive charge in this region. Furthermore, solvent molecules align so that their dipole opposes the electric field generated by the electrode.

Consider an electrode (e.g. platinum) in contact with an electrolyte solution (e.g. water). Now, also consider something impossible. Imagine that we pull the electrode and the solution apart so that there is a perfect vacuum between the inner Helmholtz plane and the electrode surface. Imagine that the solution 'surface' stays intact and appears as a static wall of positive charge and solvent molecules with their dipoles aligned.

We will examine what happens to the energy of an electron as we move it from the midpoint of a large expanse of vacuum between the solution (IHP) and to the electrode, with the horizontal axis as distance and the vertical axis energy.

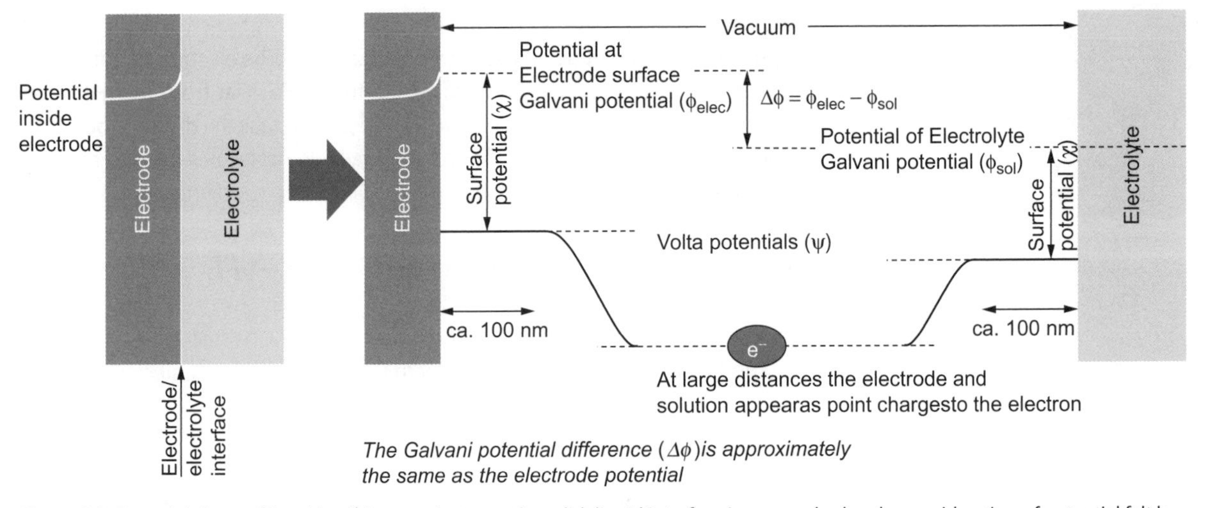

Figure 4.2 Potential change (Fermi level) in a conductor as the solid-liquid interface is approached and a consideration of potential felt by an electron as it approaches either side of the interface where a hypothetical vacuum is created between them.

The Volta potential, also called the contact potential, reflects the difference in electron number density between two materials (e.g. gold and aluminium) when they touch. In the present case it is the contact potential between the electrode and the electrolyte.

An electron in a vacuum that is far from an electrode 'sees' the electrode as a point charge with radiating electric field lines. If the electrode is polarized negatively (high Galvani potential, ϕ_{elec}) then as the electron approaches the electrode its potential increases steadily due to repulsion. As it approaches closer to the electrode (ca. 100 nm), eventually the electric field lines begin to resemble those of a plate (i.e. run parallel in one direction) and the electrode 'appears' to be a wall of charge and the repulsion the electron experiences does not increase further until it is at the electrode. The difference in the potential energy of the electron in this region compared to the vacuum is called the Volta potential (ψ). In other words, the Volta potential is the energy needed for an electron to move from the vacuum to close to the electrode. The difference between the Galvani potential (ϕ_{elec}) and the Volta potential (ψ) is denoted χ, the surface potential. We can extend the same argument to consider an electron approaching from the vacuum to the surface of a solution (which of course is not a situation that is physically possible!), as shown in Fig. 4.2. Again, as the electron approaches the solvent wall to within 100 nm its potential changes sharply to reach the Volta potential, and thereafter does not change since the solvent wall no longer appears as a point charge.

We can now ignore the Volta potentials as in reality we do not work with vacuum's (well, in solution electrochemistry that is!). We are really only interested in the difference in the Galvani potentials at each side of the solid liquid interface that the electron has to cross; that is, the difference in the Galvani potentials of the electrode and of the solution ($\phi_{elec} - \phi_{sol} = \Delta\phi$). This potential difference is the driving force that is available to do work.

The difference in Galvani potentials can be viewed as the chemical potential available and can be adjusted easily by polarization of the electrode either positively (decrease) or negatively (increase). If a molecule is far from the IHP, then all the energy available is used to move the electron across the distance between the electrode and the molecule. Therefore, available molecules need to be at or even within the IHP (i.e. adsorbed) to make full use of the driving force.

4.3 Justification of the relation $\Delta G = -\nu FE$

Where does the relation $\Delta G = -\nu FE$ come from? In Chapter 1, we discussed this issue briefly by considering the electrical work done in an electrochemical reaction

$$W_{elec} = -\Delta E = -nF\Delta E$$

which is equal to the change in Gibbs energy for the reaction under conditions of constant temperature and pressure, and hence $\Delta G_{T,P} = -nF\Delta E$. In this section, we will walk through a justification of this relation based on (electro)chemical potentials (see Box 4.1).

When we consider the change in Gibbs energy for a reaction we should consider the chemical potential of all components in the solution, however, in an electrochemical reaction, typically only the chemical potentials of the species undergoing oxidation or reduction need to be considered.

Note that work done under electrochemical conditions does not involve a significant change in pressure or volume.

Box 4.1 Chemical (μ) versus electrochemical ($\bar{\mu}$) potential

The standard chemical potential (μ°) of a species is the Gibbs energy per mole of a pure substance. The chemical potential of a species in solution can deviate from that of a pure substance because of interactions with the solvent and other solutes. The chemical potential of species 'i' (μ_i) is the partial derivative (change) of the Gibbs energy with respect to change in the number of species i present, all else held constant;

$\left(\frac{\partial G}{\partial n_i}\right)_{T,n_j} = \mu_i$ and essentially means that the chemical potential of species i is the change in Gibbs energy when n_i (i.e. 1 mol) of species i is removed, or added, to a solution. The unit of chemical potential is therefore energy (Joules, J). For the most part, we assume that the chemical potential of the solution is a fixed value for a particular situation but local changes, for example, in pH at the electrode during voltammetry, can change the total chemical potential (i.e. Gibbs energy of the solution).

In electrochemistry, we are concerned with electrochemical potential, which is defined as the partial Gibbs energy $\left(\frac{\partial G}{\partial n_i}\right)_{T,n_j}$ at a specified electrical potential. In other words, electrochemical potential is the same as chemical potential but takes into account the work done in separating and bringing together charges also. For a neutral species, the chemical and electrochemical potentials are equal ($\mu = \bar{\mu}$).

For charged species, we take into account the electrostatic contribution to its total energy and the electrochemical potential is equal to the chemical potential plus the work done (W) to add a charge ($z\,e$, where 'z' is the unit charge on the molecule/atom and 'e' is the elementary charge) to a region of electrical potential (ϕ).

The chemical energy $(\Delta G)_{T,P} = -W_{non-expansion} = -nFE$, the electrical energy.

This can be understood by considering adding an electron to a molecule. It will cost more energy to do so if the molecule is negatively charged than if it is neutral or positively charged.

$W = ze\phi$	work done to add a charge to a region with electrical potential
$W = zF\phi$	per mole (at constant temperature and pressure)

Since the work done is the difference between the electrochemical potential and the chemical potential then $zF\phi = \bar{\mu} - \mu$ and hence $\bar{\mu} = \mu + zF\phi$

Again, the electrochemical potential of a pure phase (i.e. a solid, a pure liquid, or a gas with unit **fugacity**) is equal to is chemical potential $\bar{\mu} = \mu$. For electrons in a metal, the concentration of electrons does not vary significantly and hence activity does not change. Hence, the electrochemical potential of an electron in a metal: $\bar{\mu}_e - \mu_e - F\phi_m$.

Fugacity of a gas is the equivalent of the activity of species in solutions.

We will consider the following reaction:

$$H^+_{(aq)} + M_{(s)} \rightarrow M^+_{(aq)} + \frac{1}{2}H_2(g)$$

$\Delta_r G^o$ is the driving force for the overall reaction under standard conditions, that is, a piece of metal dipped into 1 M HCl(aq).

$$\Delta_r G^o = G_{products} - G_{reactants} = \left\{\frac{1}{2}\mu_{H_2} + \mu_{M^+}\right\} - \left\{\mu_{H^+} + \mu_M\right\} = \frac{1}{2}\mu_{H_2} - \mu_{H^+} - \mu_M + \mu_{M^+}$$

The **reaction coordinate diagram** (drawn arbitrarily) indicates that the reaction between the metal and a Brønsted acid is, in this case, spontaneous as written; the metal is undergoing oxidation and the protons undergoing reduction to liberate H_2 gas spontaneously. The rate of the reaction depends on the activation barrier, and the relative rates of the forward and reverse reactions, however, for the moment, we will consider only the overall driving force.

A reaction coordinate diagram represents how high in energy a system is as the reaction proceeds. The abscissa is typically, but not always, the bond length(s) that change the most in the reaction.

We can divide the reaction into two 'half-reactions' if we employ an electrochemical cell, (i) metal becoming the metal ion in solution and (ii) protons becoming hydrogen gas. Each of these reactions have their own driving force $\Delta_r G_L$ and $\Delta_r G_R$ and are indicated in line notation as:

$$M_{(s)}\left|M^+_{(aq)}\right|\left|H^+_{(aq)}, H_2(g)\right|Pt$$

LHS reaction: $M^+_{(aq)} + e^- \rightarrow M_{(s)} \quad \Delta_r G_L$

RHS reaction: $H^+_{(aq)} + e^- \rightarrow \frac{1}{2} H_2(g) \quad \Delta_r G_R$

The overall reaction: $H^+_{(aq)} + M_{(s)} \rightarrow M^+_{(aq)} + \frac{1}{2}H_2(g)$

has the driving force under standard conditions of $\Delta_r G^o = \Delta_r G_R - \Delta_r G_L$

M + H⁺ ⇌ M⁺ + ½H₂

$\Delta_r G$ driving force for reaction
$(\Delta_f G^\ddagger)$ activation energy for forward and $(\Delta_r G^\ddagger)$reverse reactions

G

Transition state

$(\Delta_f G^\ddagger)$

$(\Delta_r G^\ddagger)$

$\Delta_r G$

Reaction coordinate

When the reaction is carried out electrochemically (divided cell), the half-cell reaction occurring at each of the electrodes needs to be considered separately (Fig. 4.3). The change in Gibbs energies ($\Delta_r G_R$ and $\Delta_r G_L$) of the two half-reactions can be expressed as a difference in potential, however, each species is in a different region of electrical potential. A cation in a region of positive charge has a higher chemical potential that a cation in a region of zero or negative charge. In the following section, we will consider each half-cell in turn and then the whole cell. For simplicity, we will use one electron half-cell reactions where $z = +1$ and $n = 1$, and M is a metal.

The two half-cell reactions each involve the transfer of an electron, the energy of which can be varied by controlling the potentials of the electrodes and hence the driving force for each half-cell reaction {$\Delta_r G_L$ (LHS) and $\Delta_r G_R$ (RHS)} can be changed allowing the driving force for the reaction overall ($\Delta_r G = \Delta_r G_R - \Delta_r G_L$) to be changed. **In the following discussion M = metal, Sol = solvent.**

4.3.1 **Left-hand side (LHS) electrode–M/M⁺**

At the LHS electrode, the reaction is:

$$M^+ + e^- \rightarrow M(s)$$

with a driving force $\Delta_r G_L$. The driving force is the difference between the total electrochemical potential of the product ($\bar{\mu}_M$) and the reactants ($\bar{\mu}_{M^+} + \bar{\mu}_{e^-}$). The

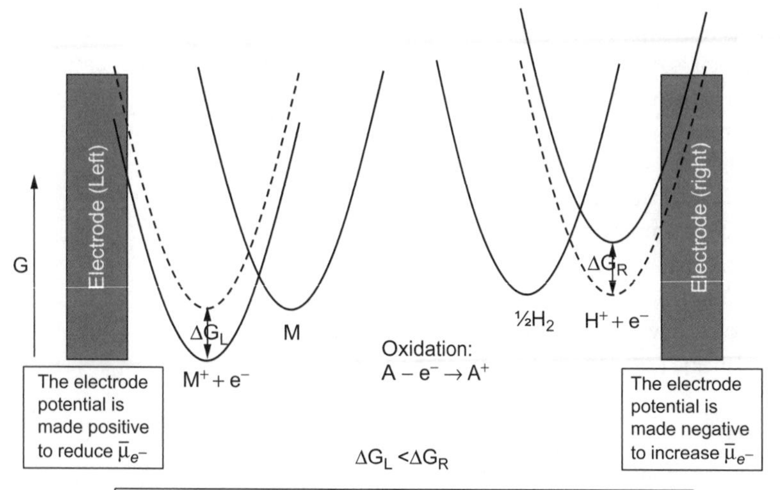

Figure 4.3 The potential energy curves for each of the half-cell reactions. On the left the metal is dissolving as the metal ion and on the right H^+ is reduced to H_2. The electrodes are drawn solely to remind us that the half-reactions are independent and occur at different places, but we should not forget that the reactions are in each case (microscopically) reversible.

electrochemical potential of the metal ($\bar{\mu}_M$) is equal to its chemical potential (μ_M) (see Box 4.1); that is, $\bar{\mu}_M = \mu_M$.

The electrochemical potential of the metal ion is its chemical potential plus the Galvani potential (ϕ) of the solvent ($\bar{\mu}_{M^+} = \mu_{M^+} + F\,\phi_{sol}$) and the electrochemical potential of the electron is its chemical potential plus the Galvani potential of the electrode ($\bar{\mu}_{e^-} = \mu_{e^-} - F\,\phi_M$). Therefore:

For simplicity we take $z = +1$ for the metal ion and -1 for the electron. The electron has a -1 charge and hence it is '$-F\Delta\phi$'.

$$\Delta_r G_L = \bar{\mu}_M - (\bar{\mu}_{M^+} + \bar{\mu}_{e^-}) = \mu_M - (\mu_{M^+} + F\,\phi_{sol} + \mu_{e^-} - F\,\phi_M) =$$
$$\mu_M - \mu_{M^+} - \mu_{e^-} + F\,(\phi_M - \phi_{sol})$$

Since the Galvanic potential difference at the LHS electrode is $\phi_M - \phi_{sol} = \Delta\phi_L$ then $\Delta_r G_L = \mu_M - \mu_{M^+} - \mu_{e^-} + F\,\Delta\phi_L$

4.3.2 **Right-hand side (RHS) electrode–H^+/½H_2**

We now consider the RHS electrode reaction, the reduction of protons to hydrogen gas.

$$H^+ + e^- \rightarrow ½\,H_2$$

The driving force for the reaction is ΔG_R, which is (half) the electrochemical potential of H_2 less the electrochemical potential of a H^+ and an electron.

$$\Delta_r G_R = ½\,\bar{\mu}_{H_2} - (\bar{\mu}_{H^+} + \bar{\mu}_{e^-})$$

The electrochemical and chemical potentials for H_2 are identical since it is a neutral species ($\bar{\mu}_{H_2} = \mu_{H_2}$). The electrochemical potential of H^+ is its chemical potential plus the Galvani potential of the solvent ($\bar{\mu}_{H^+} = \mu_{H^+} + F\,\phi_{sol}$) and the electrochemical potential of the electron is its chemical potential plus the Galvani potential of the electrode ($\bar{\mu}_{e^-} = \mu_{e^-} - F\,\phi_{elec}$). Therefore:

$$\Delta_r G_R = ½\,\bar{\mu}_{H_2} - (\bar{\mu}_{H^+} + \bar{\mu}_{e^-}) = \frac{1}{2}\mu_{H_2} - (\mu_{H^+} + F\,\phi_{sol} + \mu_{e^-} - F\,\phi_{elec})$$
$$= \frac{1}{2}\mu_{H_2} - \mu_{H^+} - \mu_{e^-} + F\,(\phi_{elec} - \phi_{sol})$$

Since the Galvani potential difference of the right–hand side is $\phi_{elec} - \phi_{sol} = \Delta\phi_{RHS}$:

Then $\Delta G_R = ½\,\mu_{H_2} - \mu_{H^+} - \mu_{e^-} + F\,\Delta\phi_{RHS}$

Hence, the electrochemical potentials are separated into their chemical potential terms and charge (electrostatics).

4.3.3 **Complete cell**

The difference between the equations for the left- and right-hand side reactions:

$$\Delta_r G_R = ½\,\mu_{H_2} - \mu_{H^+} - \mu_{e^-} + F\,\Delta\phi_{RHS}$$

$$\Delta_r G_L = \mu_M - \mu_{M^+} - \mu_{e^-} + F\,\Delta\phi_{LHS}$$

$\Delta_r G_R - \Delta_r G_L$ yields the driving force for the overall reaction, $\Delta_r G_{cell}$:

$$\Delta_r G_{cell} = \Delta G_R - \Delta G_L = \left\{\frac{1}{2}\mu_{H_2} - \mu_{H^+} - \mu_{e^-} + F\,\Delta\phi_{RHS}\right\} - \left\{\mu_M - \mu_{M^+} - \mu_{e^-} + F\Delta\phi_{LHS}\right\}$$

$$\text{remove the brackets:} = \frac{1}{2}\mu_{H_2} - \mu_{H^+} - \mu_{e^-} + F\,\Delta\phi_{RHS} - \mu_M + \mu_{M^+} + \mu_{e^-} - F\,\Delta\phi_{LHS}$$

$$\text{combine like terms:} == \left(\frac{1}{2}\mu_{H_2} + \mu_{M^+}\right) - (\mu_{H^+} + \mu_M) + (F\,\Delta\phi_{RHS} - F\,\Delta\phi_{LHS})$$

$$\text{and since: } \Delta_r G^o = \left(\frac{1}{2}\mu_{H_2} + \mu_{M^+}\right) - (\mu_{H^+} - \mu_M)$$

$$\Delta_r G_{cell} = \Delta_r G^o + F\,(\Delta\phi_{RHS} - \Delta\phi_{LHS})$$

where the first term $\Delta_r G^o$ is the driving force for the direct reaction between M and H^+, and the second term, $F(\Delta\phi_{RHS} - \Delta\phi_{LHS})$, is the contribution from electrode polarization.

When the cell is balanced by an external power source such that the system is at equilibrium (no net reaction) and $\Delta_r G_{cell} = 0$ then:

$$0 = \Delta_r G^o + F\,(\Delta\phi_{RHS} - \Delta\phi_{LHS})$$

$$\text{hence: } \Delta_r G^o = -F\,(\Delta\phi_{RHS} - \Delta\phi_{LHS})$$

Since the Galvani potential differences are essentially equivalent to the electrode potentials (E) then:

$$E_R - E_L = \Delta\phi_{RHS} - \Delta\phi_{LHS} \text{ and therefore } \Delta_r G^o = -F\Delta E$$

or more precisely where the number of electrons (v, see Section 4.1) involved per mole of reaction is not 1 (i.e. for $Ca(s) + 2H^+(aq) \rightarrow Ca^{2+}(aq) + H_2(g)$)

$$\Delta_r G^o = -vFE$$

In summary, $\Delta_r G$ for a reaction depends on the chemical potential of the electron, which can be controlled by changing the applied potential and hence the spontaneity of a reaction is dependent on the applied overpotentials. But how fast are these reactions and what happens as we apply increasing overpotentials?

4.4 Kinetics of electrochemical reactions

The key step in electrochemical reactions is the transfer of an electron across an interface and we can use the thermodynamics of this process to predict equilibria. However, the rates at which equilibria are re-established after a system is perturbed (change in electrode potential) determine the outcome of voltammetric experiments, such as cyclic voltammetry, and, for example, how much power can be drawn from a battery. The overall rate of the reactions occurring in an electrochemical cell is manifested in the current measured and we must not forget that current can be limited by any one of the steps involved in electrode reactions including:

i) heterogeneous electron transfer kinetics
ii) chemical reactions

iii) reorganization (e.g. of solvent, ions, see Chapter 2)

iv) adsorption/desorption of species at the electrode

v) mass transport: diffusion (most relevant), migration, and convection

In this section, we will show how these rates are influenced by electrode potential, when controlled by electrode kinetics, that is, when only electron transfer kinetics (and not diffusion) is rate determining. We will make use of transition state theory and the Eyring–Polanyi equation, which describes the exponential dependence of the reaction rate constant (and hence the current) on the driving force available, to develop the Butler-Volmer equation. From this equation, we will derive the well-known Tafel equation.

4.4.1 **Electrochemical rates**

Although we refer to the movement of charge in a circuit as current (i), the current passing at an electrode is typically proportional to its area (A); that is, if the electrode is doubled in area, the current will double also. This dependence adds an extra complication in derivations and in comparing systems and hence it is often more convenient to discuss current density (j in A cm^{-2}) instead.

An electrode's **electrochemical area** is not the same as its geometric area unless it is atomically flat, which is usually not the case. As an analogy, consider papering over the entire surface of Switzerland and the Netherlands—in the former case the landscape is certainly not flat and the area that you need to cover is much more than that estimated from a its area on a map.

Regardless of the potential, there is a continuous exchange of electrons between the electrode and redox active species in solution and the net rate of oxidation/reduction at an electrode is manifested in the net current density 'j': the net result of the anodic and cathodic current densities (j_a and j_c, respectively). The net current density, therefore, is: $j = j_a - j_c$.

Be aware that we also use j and i later in our discussion of impedance spectroscopy to represent imaginary numbers in complex planes.

The net number of electrons transferred between the electrode and the solution per unit time is directly proportional to the amount of the species in solution that is oxidized or reduced (the Faraday law). These rates (with units 'M s^{-1}') can be related to the concentrations **at the electrode** (i.e. at distance $x = 0$ from the electrode) by:

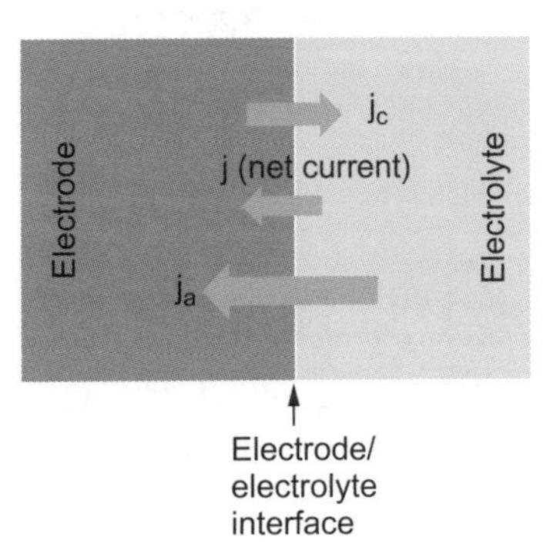

$$\text{Rate of reduction} = \frac{d[O]}{dt} = k_c [O]_{x=0}$$

$$\text{Rate of oxidation} = \frac{d[R]}{dt} = k_a [R]_{x=0}$$

With k_a and k_c, the rate constants (cm s^{-1} rather than s^{-1}, we will discuss this in later sections) for the transfer of electrons to and from the electrode, respectively, and $[R]_{x=0}$ and $[O]_{x=0}$ are the concentrations of the reduced and oxidized forms at the electrode, respectively. At equilibrium, both rates are equal (no net current flow) and the flux of electrons in each direction is referred to as the exchange velocity or rate. The current densities (j_c and j_a) are related to the reaction rates through Faraday's law and constant (F in C mol^{-1}); that is, the rate of change in concentration of the oxidized form is proportional to the number of moles of electrons ($\nu F \frac{d[O]}{dt}$), and hence the current density is

The rate of the redox reactions at the electrode are similar to simple reactions in solution with the exception that we must also consider the distance from the electrode where these reactions take place is limited by how far an electron can tunnel; ~1 nm.

$$j_c = \nu F \frac{d[O]}{dt} = \nu F k_c [O]_{x=0}$$

and similarly, $j_a = \nu F k_a [R]_{x=0}$

where ν is the number of electrons involved in each reaction.

Since both oxidation and reduction occurs simultaneously at the electrode the net current density (i.e. the current density measured, j) is given by:

$$j = j_a - j_c = \nu F k_a [R] - \nu F k_c [O]$$

Note that, although j_a, j_c have a magnitude and are positive numbers, the **net current density** (j) is defined as being positive when the net reaction is anodic and negative when the net reaction is cathodic.

In Section 4.4.2, we will use transition state theory to relate these rates to thermodynamic parameters.

4.4.2 Transition state theory and the rate of heterogeneous electron transfer

If we consider the reduction of a soluble metal ion or molecule at an electrode, then the first step is diffusion of the species towards the electrode. If the metal ion is reduced to the zero oxidation state then it must also desolvate as it breaks through the double layer to be sufficiently close to the electrode to be able to take up electrons and deposit on the electrode surface. In the case of a molecule or an ion that is not reduced to its zero oxidation state, it only has to approach the Helmholtz plane and, following electron transfer (oxidation or reduction) and a change in solvation (due to change in charge), diffuse away from the electrode. For a molecule or ion specifically adsorbed on the electrode, oxidation and reduction will still result in a change in charge and hence solvation. Hence, although we think only of electron transfer usually, there are other changes that occur that either results in a gain or loss of energy, in addition to the change in structure (bond lengths/angles, etc.) that redox reactions involve.

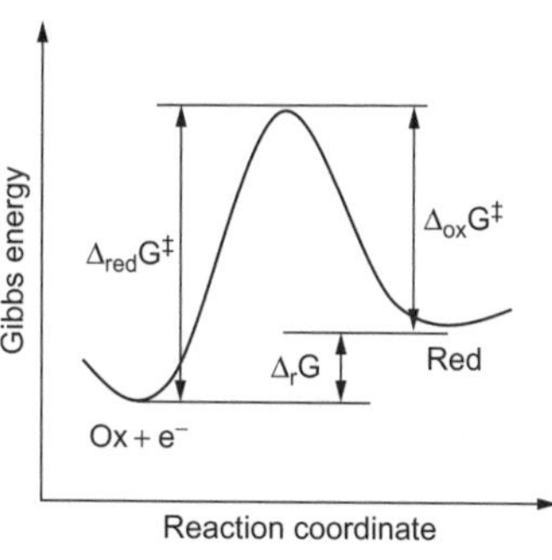

Each step/reaction is, in principle a process that requires activation, and hence the species involved must undergo an increase in its Gibbs energy, by at least $\Delta G^‡$. Since the electron transfer itself is rapid and reversible (Born–Oppenheimer approximation) the additional changes in structure that accompany electron transfer can be considered as a single elementary step and define the reaction coordinate in a reaction coordinate diagram. Transition state theory and the Eyring–Polanyi equation will be used in this section to calculate the rate constant for electron transfer to (k_a) and from (k_c) the electrode.

if you are unfamiliar with the concept of **reaction coordinate diagrams**, then review the topic in a general physical chemistry text book, for example, chapter 7 of *Modern Physical Organic Chemistry* by Ansyln and Doherty.

In Fig. 4.4, the vertical axis describes the Gibbs energy of the system. The difference in Gibbs energy (Δ_rG) between the {reduced species (R)} and the {oxidized species (O) + electron} is related to electrochemical potential ($\Delta_rG = -nFE$). The reaction coordinate represents the lowest energy path (change in structure) required to convert the reduced form to the oxidized form (and vice versa). As the (microscopic) reaction proceeds, there is an initial increase in Gibbs energy referred to as the activation energy ($\Delta G^‡$).

A detailed treatment of derivation of the Eyring–Polanyi equation, which is based on consideration of Fig. 4.4, can be found elsewhere, and in this section we will consider only the implications of the theory in the context of electrochemistry. Furthermore, the discussion is simplified by neglecting work done to bring the reactants together and for them to diffuse apart.

The rate constant for the reaction $O + e^- \rightarrow R$ can be calculated:

$$k = B_c \exp(-\Delta G^‡/k_BT) \text{ where } B_c = (\kappa k_BT/h)$$

where κ is the transmission coefficient (typically taken as being 1), k_B is the Boltzmann constant ($J\,K^{-1}$), h is the Planck constant (J s), and T is temperature in Kelvin

(K), which means that B_c has units of s^{-1}, however, since the reaction is heterogeneous; that is, the electron has to move a finite distance then it can be expressed in units of cm s^{-1}. As the electrode reaction is (microscopically) reversible there is a forward (k_c) and a reverse (k_a) reaction with activation barriers of $\Delta_{red}G^{\ddagger}$ and $\Delta_{ox}G^{\ddagger}$, respectively. Hence:

$$k_c = B_c \exp(-\Delta_{red}G^{\ddagger}/RT) \text{ and } k_a = B_a \exp(-\Delta_{ox}G^{\ddagger}/RT)$$

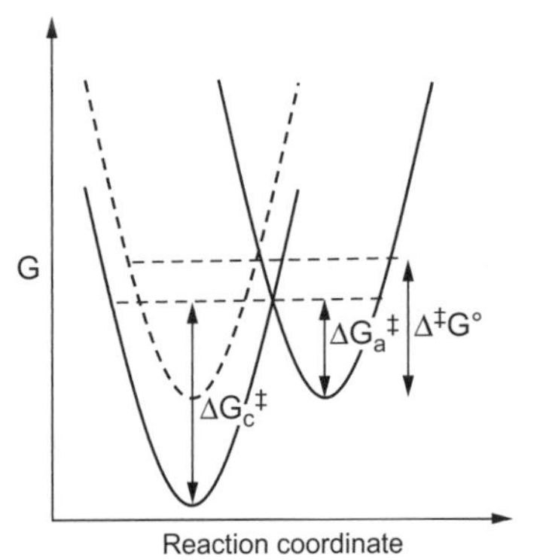

Figure 4.4 It is not possible to determine $\Delta G_a^{\ddagger}$ and $\Delta G_c^{\ddagger}$ but the Gibbs energy of activation at the equilibrium potential is $\Delta_r G^{\ddagger}$ can be determined. If the system is at equilibrium ($\Delta_r G = 0$), $\Delta G_a^{\ddagger} = \Delta G_c^{\ddagger} = \Delta^{\ddagger}G^{o}$ then $k_a = k_c$ and $j_a = j_c$ and hence $j = 0$; that is, there is no net current flow.

The electrochemical potential of the oxidized and reduced forms of the species in solution are essentially constant (assuming that the bulk solvent is not changed, e.g. from dichloromethane to methanol). In contrast, the electron's electrochemical potential is not fixed but can be changed at will by changing the electrical potential (polarization) of the electrode. Hence, although the reaction coordinate diagram drawn in Fig. 4.4 shows a situation in which the oxidation is spontaneous, by raising the energy of the electron, we can change the outcome of the reaction as shown next. To appreciate the consequences of this more deeply, we will take a step back and consider Rudy Marcus' approach to constructing **reaction coordinate diagrams**.

A reaction coordinate diagram can be drawn by considering the potential wells that describe the oxidized and reduced forms as simple parabolic curves—a parabola is a good approximation of the bottom of a potential well. By doing this we can now predict the consequence for changes in electrode (and hence electron) potential on the reaction. The shape of the parabola is determined by the normal modes of the oxidized and reduced species and hence although a change in the energy of the electron (e^-) will affect the total Gibbs energy of the 'Ox + e^-' parabola, it will not affect its shape. So, as we increase the Gibbs energy of the electron (i.e. polarize the electrode increasingly negatively) the left-hand potential well is moved vertically and the cross over point between the two wells increases and moves along the reaction coordinate towards the oxidized form (c.f. the Hammond postulate). Because the curves are parabolic we can calculate easily how a change in driving force affects the barriers and hence rate constants of the forward and reverse reactions. Note that the increase in the Gibbs energy of the crossing point is only about half the increase in the Gibbs energy of the left-hand side ($M^+ + e^-$).

B can be considered as the reaction length (roughly equal to a molecular diameter), that is, if the molecule's diameter is 1 nm the electron has to move around 1 nm and hence since B is 10^{12} s^{-1} or 1 ps then the maximum rate is 1 nm ps^{-1} or 10^5 cm s^{-1}. B is related to vibrational frequencies, that is, even if the reaction has no activation barrier (the exponential term is exp(0) = 1), then the reaction will still only proceed with a maximum rate at which nuclei can move to reorganize to new bond lengths; that is, 10^5 cm s^{-1}.

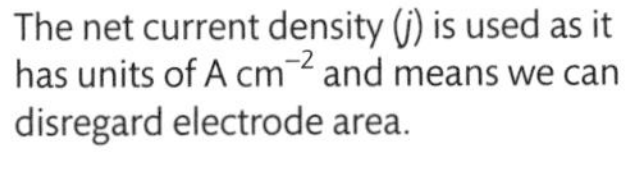

The net current density (j) is used as it has units of A cm^{-2} and means we can disregard electrode area.

In Section 4.4.3, we will discuss how the net current density (j) relates to the Galvani potential difference ($\Delta\phi$).

Our next step is to translate the visual representation of a change in electrode potential ('electron energy') to a mathematical form that we can build on. The change (Δ) in the driving force ($\Delta_r G$) for the reaction caused by the change in potential is $\Delta\Delta_r G = \nu F\eta$. Note that η is the difference in electrode potential from the equilibrium electrode potential, that is $\eta = E - E_{½}$. A change in overpotential affects the activation energies ($\Delta_{red}G^{\ddagger\prime}$ and $\Delta_{ox}G^{\ddagger\prime}$), which are increased and decreased in magnitude by a fraction (α) of the change in the driving force for the reaction ($\Delta\Delta_r G$). Hence, we calculate the Gibbs energy of activation $\Delta G^{\ddagger\prime}$ relative to the activation energy ($\Delta^{\ddagger}G^{o}$) when the overpotential is zero ($\eta = 0$) by:

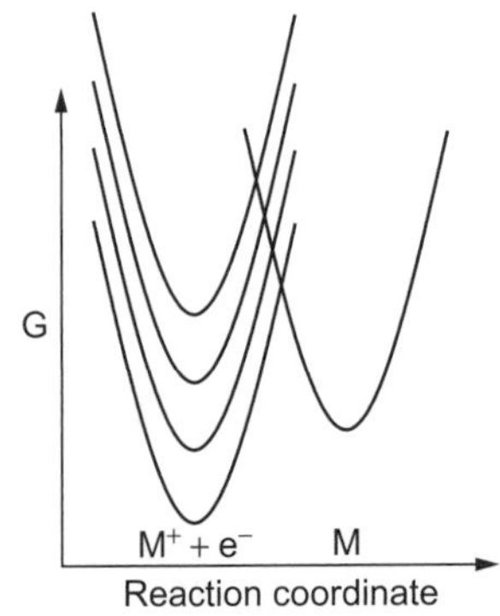

$$\Delta_{red}G^{\ddagger\prime} = \Delta^{\ddagger}G_c^{o} + \alpha\,\nu F\eta$$

$$\Delta_{ox}G^{\ddagger\prime} = \Delta^{\ddagger}G_a^{o} - (1-\alpha)\,\nu F\eta$$

***α* is called the transfer coefficient or symmetry factor** and is typically 0.5 for reversible electrochemical systems.

4.4.3 Symmetry factor (α)

The **symmetry factor/transfer coefficient (α)** is a fundamentally important parameter in electrochemistry. It can be viewed in several ways; as the symmetry of the shape of the potential wells that describe the oxidized and reduced forms of the redox active species under consideration (i.e. how close to 'mid-way' the transition state is along the reaction coordinate), or how far a species has to cross the double layer to reach the energy of the transition state (Fig. 4.5).

The **symmetry factor α** is only applicable for single elementary step reactions, while the **transfer coefficient** is used for multi–step reactions. See Bokris and Nagy, J. Chem. Ed. 1973, 50, 839, DOI: 10.1021/ed050p839 for a detailed discussion.

Regardless of its interpretation, α plays a deciding role in the response of an electrode and redox couple to a change in overpotential (η) and as a result in the net current density (j). For electrochemically reversible reactions, it has a value close to 0.5. However, when there is a large change in structure during oxidation/reduction, such as in the reduction of the enzyme cofactor NAD^+ to NADH, the potential wells of the oxidized and reduced forms are different in shape/structure and the transfer coefficient deviate strongly from 0.5.

4.4.4 The standard (or exchange) rate constant k^o

The rate constant for heterogeneous electron transfer in either direction at any overpotential can be calculated by modifying the Eyring–Polanyi equation:

In this discussion we use the Eyring–Polanyi equation, however, an analogous derivation is made elsewhere using the Arrhenius equation; $k = A\exp\left[-\frac{E_A}{RT}\right]$ and substituting E_A for $\Delta H^{\ddagger}$ and factorizing A as $A'\exp\left[-\frac{\Delta S}{R}\right]$.

$$k_c = B_c \exp\left[-\frac{\Delta_{red}G^{\ddagger}}{RT}\right] = B_c \exp\left[-\frac{\left(\Delta^{\ddagger}G_c^o + \alpha F\eta\right)}{RT}\right]$$

$$k_a = B_a \exp\left[-\frac{\Delta_{ox}G^{\ddagger}}{RT}\right] = B_a \exp\left[-\frac{\left(\Delta^{\ddagger}G_a^o - (1-\alpha)F\eta\right)}{RT}\right]$$

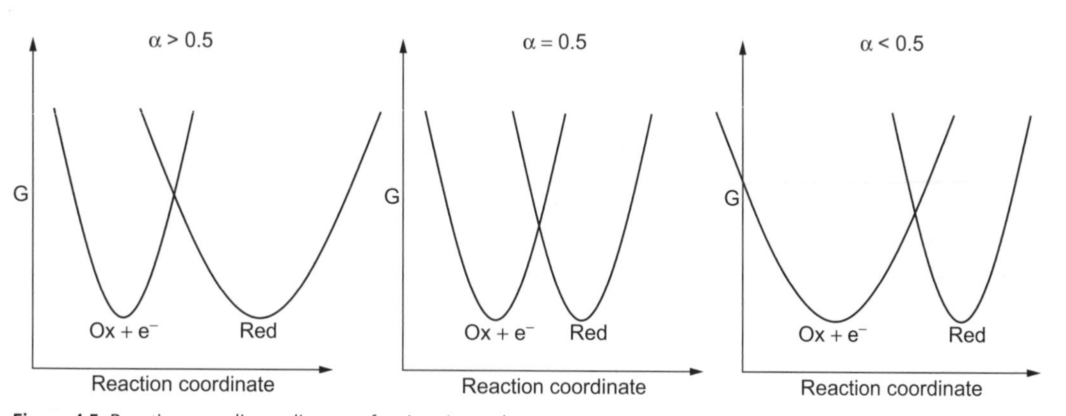

Figure 4.5 Reaction coordinate diagrams for situations where $\alpha > 0.5$, $\alpha = 0.5$, and $\alpha < 0.5$. The widths of the potential wells indicates changes in, for example, changes in bond lengths before and after oxidation.

Box 4.2 Effect of electrode surface on electron transfer kinetics

The chemical potential of the solvent (μ_{sol}) and the (electro)chemical potential of the electrode (μ_{elec}) will (almost) never be equal and hence at the solid liquid interface there is a sudden change in (electro)chemical potential ($\Delta\mu = \mu_{sol} - \mu_{elec}$). Of course, there cannot be an infinitely sharp change in potential and hence a gradient $\left(\text{potential drop}\left\{\frac{\Delta\mu}{\textit{distance}}\right\}\right)$ must exist between the potential at the electrode surface and the bulk solution. This gradient crosses the inner and outer Helmholtz planes (discussed in Chapter 3) and it can be approximated as a linear change over about 1 nm. This gradient means that electrons have to tunnel a short distance, and, as a result, a molecule has to diffuse at least to the outer Helmholtz plan in order for electron transfer to occur. Since the potential drop is over a greater distance than the thickness of the inner Helmholtz plan (ca. 0.5 nm), a species in solution does not have to 'break through' the ordered layer of ions and solvent immediately at the electrode. The requirement that a species must approach close to the electrode has a consequence for electron transfer. For example, for large molecules, in which the redox centre is buried deep within an insulating matrix (e.g. a protein), the distance to the electrode will be too great for electron transfer to occur efficiently (i.e. low rate). We will return to this problem, and a solution, when we consider the glucose sensor in Chapter 5.

Although it is tempting to treat the electrode as simply a source of electrons, that is, a passive component, it is in fact essential to understand the influence of the electrode's surface characteristics has on electron transfer kinetics. Take for example a glassy carbon electrode. Its surface can be modelled as comprising of folded ribbons of graphene/graphite, that is, sheets of sp^2 hybridized carbon atoms. However, if a glassy carbon is polarized at a sufficiently positive potential in water, such that water is oxidized to generate reactive oxygen species such as hydroxyl radicals, then the carbon atoms on the surface can undergo oxidation and over time the surface concentration of dangling CO_2H/CO_2^- groups will increase (a process called anodization). This process leads to a surface, which is essentially hydrophobic with a charge determined solely by the polarization, changing to a surface that has a permanent negative charge and hence any negatively charged species present in solution will be 'repelled' from the inner Helmholtz plan. The effect of this is to reduce the potential drop that the negatively charged species has access to (see Section 4.1.5 for our discussion of the inner and outer Helmholtz planes) and the rate of electron transfer will be decreased dramatically. This effect can be seen in the cyclic voltammograms shown in Fig. 4.6 (Box 4.1) (we will discuss cyclic voltammetry in detail in Chapter 5). In this example, a glassy carbon electrode is anodized heavily at 1.2 V in 0.1 M KNO_3 and a voltammogram recorded in a solution containing 1 mM $[Fe(III)(CN)_6]^{3-}$ in 0.1 M KNO_3. The voltammogram shows essentially no current response, meaning that electron transfer is much slower than needed to maintain a Nernstian equilibrium situation at the electrode during the cyclic voltammogram. If the electrode is subsequently polished to remove the surface layers mechanically and reveal a fresh layer of graphene like carbon free of carboxylic acid, then the expected reversible cyclic voltammogram is obtained for $Fe(III)(CN)_6]^{3-}$.

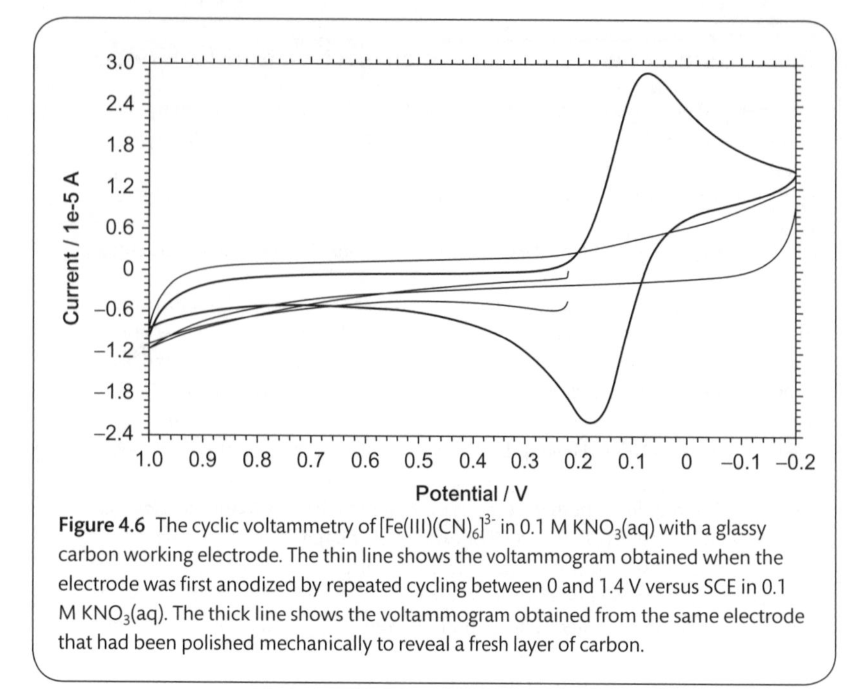

Figure 4.6 The cyclic voltammetry of $[Fe(III)(CN)_6]^{3-}$ in 0.1 M KNO_3(aq) with a glassy carbon working electrode. The thin line shows the voltammogram obtained when the electrode was first anodized by repeated cycling between 0 and 1.4 V versus SCE in 0.1 M KNO_3(aq). The thick line shows the voltammogram obtained from the same electrode that had been polished mechanically to reveal a fresh layer of carbon.

The rate constant for heterogeneous charge transfer (k^o/cm s^{-1}) is > 1 cm s^{-1} for fast reactions where $E_{½} \approx Ef^o$ and $\alpha \approx 0.5$.

Riboflavin (vitamin B12)

Nictionamide (part of NAD^+, NADH) in oxidised and reduced forms

Hence, the rate constants for the reactions in each direction are exponentially dependent on the change in electrode potential and therefore the net current density (j) observed is also exponentially dependent on applied potential.

However, the actual effect of changing potential is often limited by mass transport phenomena, which will be discussed in Chapter 5. Spatial consideration of the electrochemical reactions can begin with a simple model in which the reactant (oxidized form) diffuses to the electrode and undergoes reduction to the reduced form. The reduced form thereafter diffuses away from the electrode.

Consider the equation describing the current (cathodic) that flows when such a reaction can take place: $j_c = \nu Fk_c[O]_{x=o}$, where $[O]_{x=o}$ is the concentration of the oxidized form at the electrode. Although the current increases as the rate constant increases, the concentration of the oxidized form at the electrode will limit the current observed to the rate at which the oxidized form can reach the electrode by diffusion (mass transport limited). In Section 4.4.5, we will disregard this experimental aspect and assume that the rate at any overpotential is governed by electron transfer kinetics.

4.4.5 Dependence of current on overpotential–The Butler-Volmer equation

Combining the current densities in each direction (to and from the electrode):

$$j = j_a - j_c = \nu F\ k_a\ [R] - \nu F\ k_c\ [O]$$

and replacing k_a and k_c (see Section 4.4.4) yields the Butler–Volmer formulation:

$$j = vFB_a[R]\exp\left[-\frac{\Delta_{ox}G^{\ddagger}}{RT}\right] - vFB_c[O]\exp\left[-\frac{\Delta_{red}G^{\ddagger}}{RT}\right]$$

which can be used to predict the net current as a function of overpotential.

If we express $\Delta_{ox}G^{\ddagger}$ and $\Delta_{red}G^{\ddagger}$ in terms of overpotential and activation energy at the equilibrium potential (see Section 4.4.4), then:

$$j_a = vFB_a[R]\exp\left[-\frac{(\Delta^{\ddagger}G_a^o - (1-\alpha)F\eta)}{RT}\right] = vFB_a[R]\exp\left[-\frac{\Delta^{\ddagger}G_a^o}{RT}\right]\exp\left[\frac{(1-\alpha)F\eta}{RT}\right]$$

and:

$$j_c = vFB_c[O]\exp\left[-\frac{(\Delta^{\ddagger}G_c^o + \alpha F\eta)}{RT}\right] = vFB_c[O]\exp\left[-\frac{\Delta^{\ddagger}G_c^o}{RT}\right]\exp\left[-\frac{\alpha F\eta}{RT}\right]$$

When the overpotential (η) is zero (i.e. at the equilibrium potential $E_{1/2}^{o}$), the last term in each equation becomes exp(0) and hence '1' and the magnitude of j_a is equal to j_c. The net current j is therefore zero. At this potential $j_a = j_c = j_o$ (the exchange current density):

$$j_o = vFB_a[R]\exp\left[-\frac{\Delta^{\ddagger}G_a^o}{RT}\right] = vFB_c[O]\exp\left[-\frac{\Delta^{\ddagger}G_c^o}{RT}\right]$$

if $\eta = 0$, then $\exp\left[\frac{(1-\alpha)F\eta}{RT}\right] = \exp\left[\frac{(1-\alpha)F(0)}{RT}\right] = \exp 0 = 1$.

Hence the equations can be written as:

$$j_a = j_o\exp\left[\frac{(1-\alpha)F\eta}{RT}\right] \text{ and } j_c = j_o\exp\left[-\frac{\alpha F\eta}{RT}\right]$$

and overall:

$$j = j_o\exp\left[\frac{(1-\alpha)F\eta}{RT}\right] - j_o\exp\left[-\frac{\alpha F\eta}{RT}\right]$$

Reactant (Ox) Product (Red)
Ox + e⁻ → Red
Electrode

The equilibrium potential $E_{½}^{o}$ is defined as the potential at which the anodic and cathodic current densities are equal in magnitude and if $j_c = j_a$ and $\alpha = 0.5$ then the net current $j = 0$ and the exchange current density (j_o).

The **exchange current density**, j_0 is the current per unit area passing across the solid liquid interface in each direction at the equilibrium potential.

Although the Butler–Volmer equation can be used to predict current density as a function of overpotential, experimentally this is only useful if the current is limited by j_o, that is, mass transport has to be sufficiently efficient to not limit the current by the build-up of concentration polarization. The exchange current density can vary over two orders of ten orders of magnitude; for example, in the reaction $H^+ + e^- \rightarrow H_2$, j_o is 2×10^{10} times higher at a palladium electrode than at a mercury electrode!

4.5 The Tafel plot

The plot of the natural logarithm of the absolute magnitude of the current density ($|j|$) against the overpotential (η) (i.e. ln $|j|$ vs. η, see, Fig. 4.7) is called a Tafel plot.

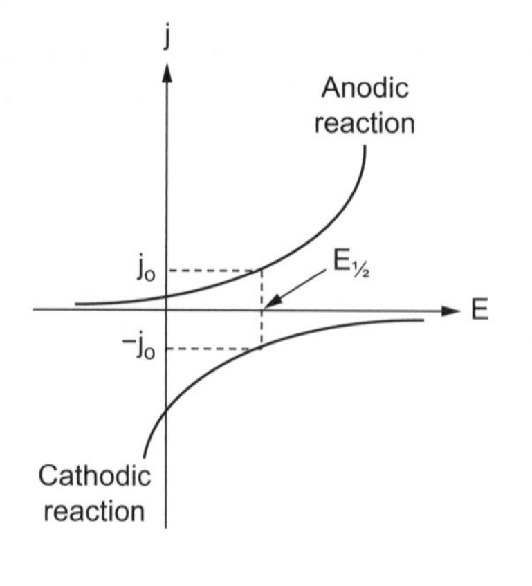

The peculiar shape of the plot can be understood using the equations developed in Section 4.4.

The net current j is the difference in electron flux in each direction:

$$j = j_a - j_c = j_o \exp\frac{(1-\alpha)F\eta}{RT} - j_o \exp\frac{(-\alpha)F\eta}{RT}$$

The equation simplifies when the overpotential ($|\eta|$) is large because one of the two terms approaches zero and that region of the Tafel plot is approximately linear with slopes of $(1-\alpha)\, F/RT$ and $\alpha\, F/RT$. Close to the equilibrium potential ($E_{1/2}^{o}$), however, the two terms are similar in magnitude and the slope changes to $|j_o F\eta|$. The three situations are considered in detail here:

At low overpotentials (close to $E_{1/2}^{o}$), when $\eta \approx 0.01$ V then $F\eta/RT << 1$ and:

when $x << 1$ then $\exp(x) = 1 + x$.

$$j \approx j_o\left\{1+\frac{(1-\alpha)F\eta}{RT}\right\} - j_o\left\{1+\frac{(-\alpha)F\eta}{RT}\right\} = j_o\left\{1+\frac{(1-\alpha)F\eta}{RT}-1-\frac{(-\alpha)F\eta}{RT}\right\}$$

$$= j_o\left\{\frac{(1-\alpha)F\eta}{RT}-\frac{(-\alpha)F\eta}{RT}\right\} = j_o\{1-\alpha-(-\alpha)\}\frac{F\eta}{RT} = j_o\{1-\alpha+\alpha\}\frac{F\eta}{RT} = j_o\frac{F\eta}{RT}$$

$j \approx j_o \frac{F\eta}{RT}$; that is, the current is proportional to overpotential (Ohm's law), which is apparent when the equation is rewritten as: $\eta \approx j\left(\frac{RT}{Fj_o}\right)$

At large positive overpotentials, the second exponential term is much smaller than the first and:

$$j = j_o \exp((1-\alpha)\, F\, \eta\, /RT) - j_o \exp(-\alpha\, F\, \eta)/RT) \approx j_o \exp((1-\alpha)\, F\, \eta\, /RT)$$

which can be rearranged to the equation of a line: $ln\,|j| = ((1-\alpha)F/RT)\,\eta + ln\, j_o$.

Similarly, at negative overpotentials the second exponential term is much larger than the first term and hence:

$$j = j_o \exp((1-\alpha)\, F\eta\, /RT) - j_o \exp(-\alpha\, F\eta)/RT) \approx j_o \exp(-\alpha F\eta\, /RT)$$

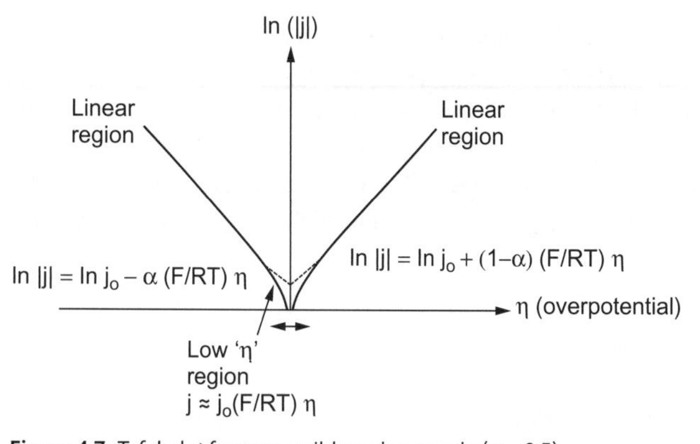

Figure 4.7 Tafel plot for a reversible redox couple ($\alpha = 0.5$).

which can be rearranged to the equation of a straight line: $ln\,|j| = (-\alpha F/RT)\,\eta + ln\,j_o$.

Consideration of the Tafel equation and the Tafel plot shown in Fig. 4.7 raises a number of key points:

- Although j_a and j_c will never be zero, if they are equal in magnitude then they will cancel and the net current (which is measured) will be zero.
- In the case of more complex reactions, for example, where there are a number of steps (with each step having a different value of v) then the value for v to be used is the number of electrons involved at the rate determining step in the reaction mechanism and not the overall number of electrons transferred. In this aspect, the Tafel analysis differs from the Nernst equation and in relating current density to the rate constant where the overall stoichiometry is taken into account.
- The Tafel equation applies for both redox active species in solution as well as on surfaces and both (e.g. also to corrosion and electroplating).
- α is typically 0.5 for a single step reaction.

Finally, it should be remembered that the Tafel plot is only valid under conditions where mass transport is faster than the rate of electron transfer. This is not usually the case, as we will see in Chapter 5, and hence care should be taken to consider carefully experimental design when applying the Tafel equation to real problems.

4.6 **Summary**

This chapter should provide you with an understanding of:

- the concept of heterogeneous electron transfer between two phases (solid/liquid)
- the relation between chemical and electrochemical potential, electrode potential, and overpotential
- the origin of the relation between potential and standard Gibbs energy change for a reaction
- calculation of current through consideration of electron transfer rates and Eyring–Marcus Theory
- the derivation of the Butler–Volmer equation
- the derivation of the Tafel equation and plot, and determination of exchange current density

4.7 **Exercises**

4.1. Explain why the standard rate of electron transfer is lower for NADH oxidation than for riboflavin.

4.2. Using a spreadsheet and the Butler–Volmer equation, predict the anodic and cathodic current and the net current between $\eta = -1$ V and $\eta = +1$ V and α is

0.5, 0.45, 0.35, and 0.75 where $j_o = 0.01 A cm^{-2}$. Use the following form of Butler–Volmer equation $j = j_o \exp\left[\frac{(1-\alpha)F\eta}{RT}\right] - j_o \exp\left[-\frac{\alpha F\eta}{RT}\right]$

4.3. Show how the Butler–Volmer equation $\left(j = j_o \exp\left[\frac{(1-\alpha)F\eta}{RT}\right] - j_o \exp\left[-\frac{\alpha F\eta}{RT}\right]\right)$ can reduce to the Nernst equation when $\eta \approx 0.01$ V then $F\eta/RT << 1$; Verify this statement. Hint: remember that current density is related to concentration at the electrode.

4.4. Use a spreadsheet or other suitable graphing program to predict the shapes of the Tafel plot for a series of situations where α is 0.5, 0.45, 0.35, and 0.75. What do you notice about how the plot's shape changes?

5 Dynamic electrochemical techniques

5.1 Introduction

In this chapter, we will discuss both the equipment used in electroanalytical chemistry and look closely at the processes that occur at the **electrode**/electrolyte interface during dynamic experiments (i.e. experiments in which the extent of electrode polarization ($\Delta\phi$) changes, either suddenly, or continuously over time). We will build on the concepts of the electrode solution interface, mass transfer by diffusion, heterogeneous electron transfer, and electrode polarization/depolarization developed in Chapter 4. In particular, we will focus on the development of concentration gradients (called depletion or Nernst diffusion layers) through the stagnant layer to the bulk solution (Section 3.6). A number of general concepts relevant to voltammetry will be discussed before going deeper into describing the time and potential dependent processes that occur at the working electrode when the electrode potential changes. In this discussion, we introduce several key equations that are ubiquitous in the electrochemical literature, including the Cottrell and Randles–Sevĉik equations. The discussion continues by briefly examining the most commonly encountered electroanalytical experiments in teaching and research laboratories.

The depth of the stagnant layer depends on conditions and is seldom well defined even when convection is forced by rapid stirring, rotating the electrode or pumping of solution across the electrode. In this chapter we will assume that the stagnant layer is well defined in order to simplify the explanation, however, as discussed in section 3.8 is is never really the case.

5.2 Dynamic electrochemistry

Electroanalytical chemistry relies heavily on voltammetric techniques in which the potential at an indicator (working) electrode, which is polarizable, is controlled relative to a fixed (non-polarizable) reference electrode. The current that results from a change in the potential (polarization) of a working electrode is recorded. These techniques yield plots called voltammograms, for example, Fig. 5.1 (and called polarograms when the working electrode is mercury). In contrast to potentiometric sensing where the goal is to avoid disturbing the system (i.e. avoid the flow of current), with voltammetric experiments we measure the current that flows when species in solution depolarize (i.e. work against our efforts to polarize) the working electrode, and hence chemistry happens at, at least, two electrodes.

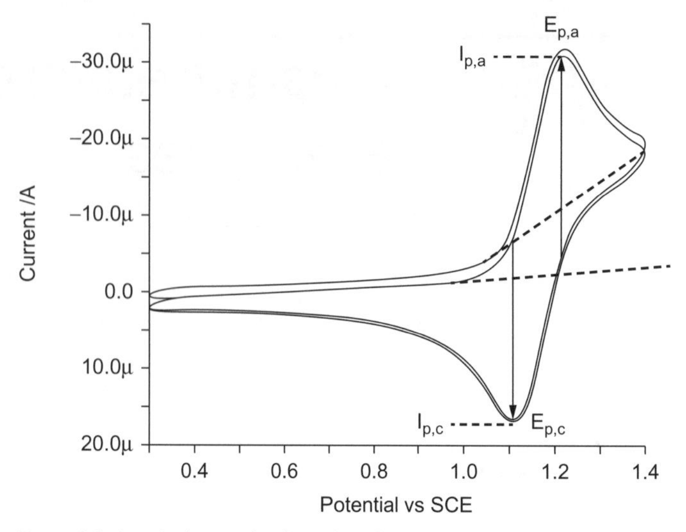

Figure 5.1 A typical example of a cyclic voltammogram of a species in solution showing the characteristic peak currents ($I_{p,a}$, and $I_{p,c}$) and potentials ($E_{p,a}$, and $E_{p,c}$) at which these currents are reached.

Mercury, the only liquid metallic element, was once a mainstay of electrochemistry and although it is still, its toxicity and environmental impact has led it to being phased out in most areas and replaced by, for example, glassy carbon electrodes. It is now a specialist electrode material and since its use is a field in itself (polarography), it will not be considered in this book.

In other fields, especially in photovoltaics, so-called *I-V* plots are used also, however, in these plots the current does not depend on scan rate.

In dynamic electrochemistry we work under conditions where current can flow and hence either the potential or the current has to be controlled. We control the potential/current (potentiostatic/galvanostatic control, respectively) using an external variable power supply with which we can actively build-up a positive (positive polarization) or negative potential difference at the surface of the working electrode; in Chapter 4 this was described as lowering and raising the Galvani potential of the electrode, respectively.

Polarization of an electrode in contact with a solution will be counteracted (i.e. the electrode is depolarized) to a degree by solvent/electrolyte reorganization (the double layer capacitance, C_{dl}), and hence a non-Faradaic current will flow transiently if the electrode potential is changed. In addition, specifically adsorbed species (i.e. molecules covalently or non-covalently attached to the electrode) can desorb and vice versa. As discussed in Section 3.1, the IHP, OHP and diffuse layer are together only a few nanometres thick. A somewhat thicker layer of solvent, the stagnant layer (see Section 3.6 and 5.1), is also important in our discussion. Within this layer, the concentration gradients build-up during experiments to form the Nernst diffusion layer. The Nernst diffusion layer is a central consideration in our interpretation of the I/V curves we generate through techniques such as linear sweep and cyclic voltammetry.

5.2.1 **General concepts in voltammetry**

Several concepts encountered regularly in gathering and interpreting electrochemical data including (i) Faradaic and non-Faradaic current, (ii) the three-electrode arrangement, and (iii) the impact of cell resistance on voltammetric experiments. In this section, we will comment on these issues in turn.

5.2.1.1 **Faradaic current versus non-Faradaic current**

Although an ammeter is ignorant of the origin of the current that flows in a circuit, we make a conceptual distinction between; (i) current generated by processes that do not involve electron transfer between the electrode and solution (non-Faradaic), such as current due to double layer charging, and (ii) by processes that involve electron transfer (Faradaic), that is, redox reactions, to and from the electrode and solution.

If the potential at an ideal polarizable electrode is changed suddenly from, for example, the **open circuit potential (OCP)**, by +100 mV then a movement of charge (ions and solvent dipole reorientation), and the change in equilibrium position between oxidized and reduced forms of a redox active species on the surface, will induce a current to flow in the circuit. If there are no redox active species in solution that can counter the change in electrode polarization, then a current due to solvent reorganization will flow transiently and after a short time return to zero. In contrast, if a redox active species that can depolarize the electrode (can be oxidized or reduced) is present in solution, then an additional current will flow for a longer time (Fig. 5.2). This current depends on the net rate of diffusion of the redox active species to the electrode and the overpotential applied. This latter process is discussed in detail in Section 5.2.1.2.

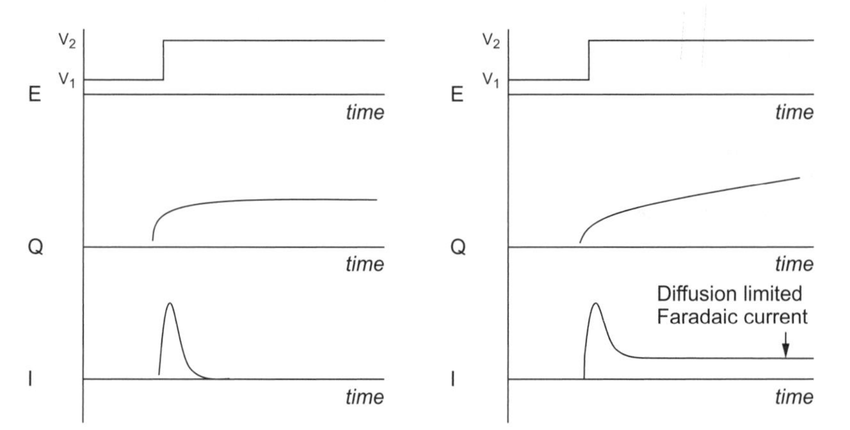

Figure 5.2 Time dependence of charge passed (Q) and current (A) following a potential step (left) in the absence and (right) in the presence of a redox active species in solution.

5.2.1.2 **The three-electrode arrangement**

Dynamic (i.e. current flowing) electrochemistry can be carried out with two electrodes; a **'reference' electrode (RE)** (that behaves as close as possible to that of an ideally non-polarizable electrode) and a **'working'** or indicator **electrode (WE)** (that is usually close to ideally polarizable in its behaviour). However, normally we use a three-electrode arrangement, which includes a third **'counter' electrode (CE).**

Other arrangements include a second WE (i.e. a four-electrode arrangement) but we will not discuss these here.

The benefit of the three-electrode arrangement is that, although close to ideal, if a sufficiently large current is drawn through the reference electrode it will eventually undergo some polarization (and hence its potential will 'float'). In this situation, the potential at the WE will not be well-defined either. Furthermore, the potential difference between the electrodes cannot be measured accurately if significant current is flowing between them. A CE is used to solve this problem and allows current to flow from the WE to the CE and vice versa with the result that the current flowing between the WE and RE is negligible. In this way, the current passing via the WE can be measured by placing an ammeter between the WE and CE and the potential of the WE relative to the reference electrode is measured by placing a voltmeter between the RE and WE as shown in Fig. 5.3.

Since current does not pass through the reference electrode, its composition, and hence potential, will not change over time. A key point to realize, however, is that we can only change the potential of the WE relative to the CE, which is achieved by adjusting a variable power supply positioned in the circuit between the WE and CE. We vary the potential difference between the WE and CE as much as is necessary (up to 20 V or higher!) until the polarization of the WE relative to the reference electrode is that which we require. Hence we only sense (measure) the potential difference between the WE and RE, and if we want to change this value then we do so by actively changing the potential difference between the WE and CE.

The current that we measure, that is the current that passes through the ammeter, is usually limited by one component/reaction in the circuit; one process in the system is the slowest and determines the overall rate/current. It is essential that this 'rate limiting process' is a process at the WE. This requirement is fulfilled by using a CE with a surface area that is much larger than that of the WE. For example, a 1 mm diameter WE (e.g. a Pt button electrode) would be used with a 4 cm^2 Pt sheet CE. Importantly, the reactions that lead to depolarization of

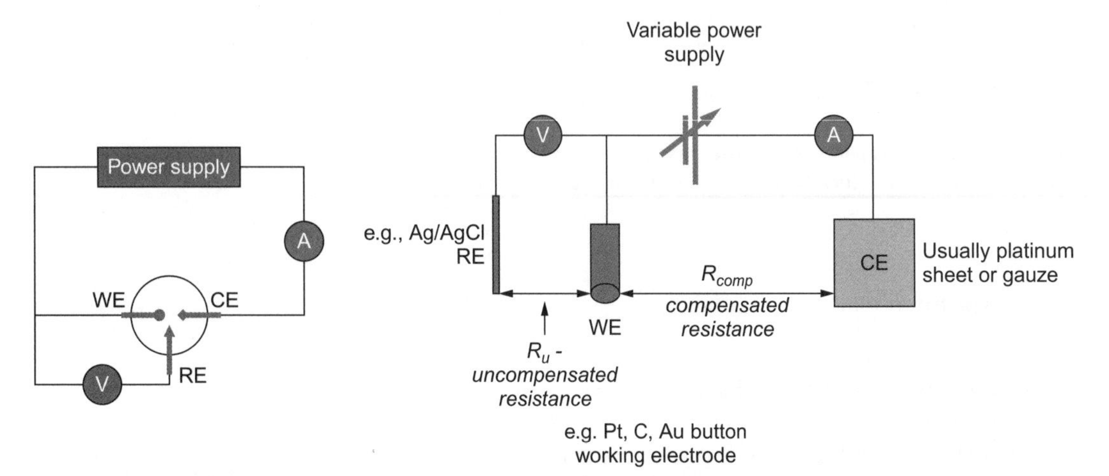

Figure 5.3 (left) Simplified circuit for a three-electrode cell, and (right) diagram showing the electrodes and circuit as well as compensated and uncompensated resistance

the CE (typically reactions such as $H^+ + e^- \rightarrow \frac{1}{2} H_2(g)$) are not usually a concern; we generally don't even think about them. However, it is essential that the depolarization is rapid compared to the reactions occurring at the WE and also that the reactions do not lead to passivation of the CE (i.e. formation of a polymer film that blocks electron transfer).

5.2.1.3 Cell resistance/conductivity

Electrochemistry relies on the conduction of electrons through a circuit and ions through a solution. If the solution is not able to conduct ions then current will not flow and the **impedance** (resistance) will cause energy (and potential) to be dissipated as heat. In a poorly conducting solvent the resistance will be high and result in a large '***iR* drop**' (or iR_{drop}), which must be compensated for electronically. A dissociable salt is added to increase the ionic strength of the solvent to overcome this. The salt must therefore be fully, or at least largely, dissociated in the solvent.

In a two-electrode system, the Galvani potential difference between one electrode and the solution ($\Delta\phi$) is accompanied by a second Galvani potential difference between the solution and the second electrode. As discussed in Chapter 4, the net result (the potential difference measured) is the difference in the difference of the Galvani potentials ($\Delta\Delta\phi$). In practice, we cannot ignore the extra 'electronic components' in the circuit. For example, the solution behaves like a resistor in series, and its contribution to the voltage drop in the circuit depends on the current. It is referred to as the iR_{drop} between electrodes.

The only energy available to do work (transfer an electron) is the difference in Galvani potentials ($\Delta\Delta\phi$). If the potential difference we apply between two electrodes is, for example, 2 V, but the iR_{drop} accounts for 1.5 V, the only energy available to do work across the inner Helmholtz plane is 0.5 V. The potential of the electrode and the potential at the inner and outer Helmholtz plane determine whether electron transfer occurs or not. Hence, the voltage that is displayed on the voltmeter in the external circuit is the sum of the work that can be done at the electrodes **and** the voltage drop due to solution resistance (iR_{drop}).

The resistance between the WE and CE depends on the resistivity of the electrolyte and their separation (distance, see Chapter 1). If current flows then there will be a potential drop across the solution with a magnitude of iR (from $V = iR$). This is not usually a problem as we simple adjust the power supply to compensate for this voltage drop and hence the resistance between the WE and CE is referred to as **compensated resistance**.

In a three-electrode arrangement, with a CE, a non-polarizable reference electrode and the WE, the potential difference that is measured/applied between the working and reference electrodes (e.g. $E_{applied} = 1.2$ V) does not equal the actual potential drop at the electrode. Instead, it includes the energy loss by uncompensated solution resistance (iR_u, also referred to as iR_{drop}), where R_u is the resistance due to the electrolyte solution between the reference electrode and the WE and 'i' is the current flowing between the working and CE, modelled as an equivalent circuit in Fig. 5.4. The resistance due to electrolyte between the WE and RE

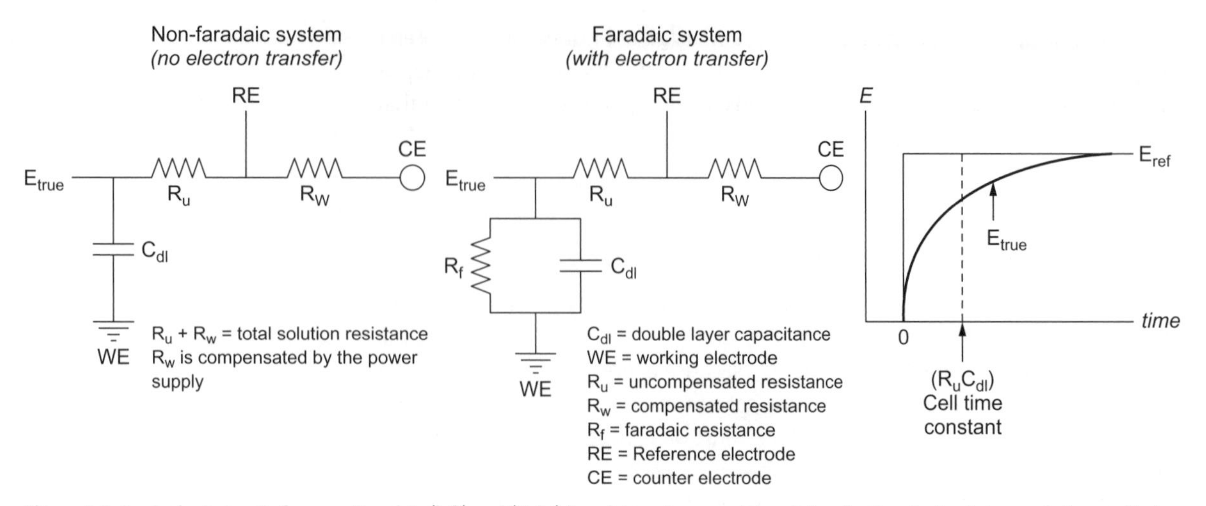

Figure 5.4 Equivalent circuits for non-Faradaic (left) and (right) Faradaic systems and time taken for the electrode to reach the applied potential.

iR_u is the product of the resistance of the solution between the WE and RE and the current between the WE and CE. This may seem counterintuitive but the current flows at the WE and hence even if it comes via the CE, the current is still flowing 'between' the WE and RE.

cannot be compensated by adjusting the power supply and hence is referred to as uncompensated resistance.

The **actual potential** (E_{true}) at a WE is therefore $E_{applied} - iR_u = E_{true}$. At low currents and low solution resistances, for example, 100 Ω and 1 μA, the difference between the potential read from the voltmeter and the potential at the WE is only 0.1 mV. If the current is 1 mA and the solution resistance 1 kΩ; however, then the potential at the electrode would be 1 V less than the value indicated by the voltmeter!

A further consideration is the effect of solution resistance on the cell time constant, which defines the shortest time domain over which the cell will respond fully to a perturbation (i.e. how fast the potential at the WE will actually change to the intended potential).

$$E_{true} = E_{applied}\left[1 - \exp\left(\frac{-t}{R_u C_{dl}}\right)\right]$$ where $R_u C_{dl}$ is the cell 'time constant'

When the double layer is charged fully, that is as 't' becomes long, the exponential term approaches zero and $E_{true} \cong E_{applied}$

The cell time constant and iR_u can be reduced by:

- decreasing the magnitude of the double layer capacitance, C_{dl}, by using a smaller electrode, for example, a micro-electrode, (remember that C_{dl} is proportional to electrode area).
- increasing solution conductivity by adding electrolyte and increasing the polarity of the solvent (i.e. decrease R_u and R_W).
- decreasing the R_u by positioning the RE close to the WE since the magnitude of the resistance is proportional to the distance between the electrodes.
- decreasing analyte concentration, and thereby Faradaic current.

As a final practical consideration, the potential difference between the WE and CE can be as much as 20 V during routine experiments especially in organic apolar solvents. In principle, the electric field gradient generated could induce mass transport by migration, however, the electrolyte shields the analyte from these field gradients provided that the ion pairs are dissociated. A more important practical concern in large scale electrosynthetic applications is temperature changes due to resistive heating.

5.2.2 **The current response to potential steps and ramps**

Up to this point, we have considered a situation at an electrode that is at a steady state (equilibrium) at a particular potential and the concentrations of species (oxidized/reduced forms) at the electrode satisfy the Nernst equation. In this section, we will consider what happens when the potential of the WE is changed either in steps or continuously and use the Nernst equation to establish the boundary conditions for the changes we can expect.

The simplest dynamic experiment that we can carry out is a potential step. In this experiment we change the polarization of the WE (e.g. by 10 mV or 100 mV, etc.) and follow over time the response of the system, manifested in the current that flows at the WE (Fig. 5.2). An alternative approach is to change the potential at a constant rate (a so-called linear sweep) and measure the current that flows as the sweep is in progress to generate a plot of current versus potential (I vs. V). If we do this with a resistor, for example, then we will obtain straight-line graph with an intercept of 0 A at 0 V and a slope that is equal to the conductance (i.e. the inverse of resistance, $1/R$). The plot obtained (I vs. V) is referred to as a voltammogram, see, for example, Fig. 5.1.

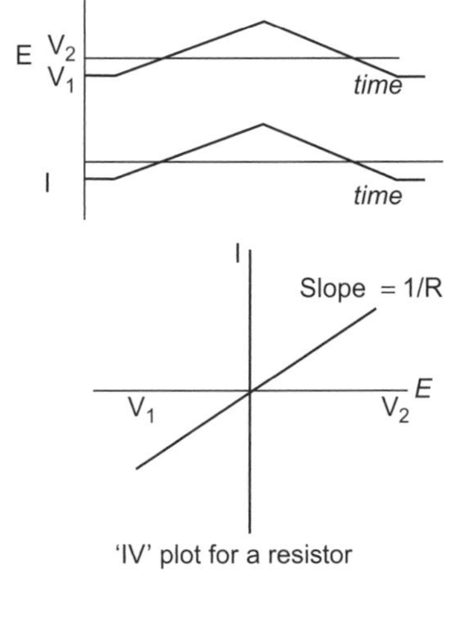

'IV' plot for a resistor

A key point to remember with voltammograms is that the current indicated at a particular potential is not the current that we would measure if we were simply to hold the electrode at that potential but instead it is the current that we measure at the moment that we are at that potential during a potential sweep. This is a major difference between solution electrochemistry and solid-state devices such as solar cells and resistors where the current is the same whether or not the potential is fixed or changing over time.

5.2.2.1 **Current response to a potential step and a potential ramp in the absence of a redox active species**

In the absence of a redox active species, or where the potentials involved are far (> 500 mV) from the formal potentials of a redox active species present, a change in electrode potential should result in little or no current. However, the solvent and electrolyte reorganization (as discussed at the start of Section 5.2) will result in the transient flow of current. Hence, following a sudden change in potential, a current will flow with a delayed/stretched response because the system takes a finite time to react (note the discussion of the RC time constant in Chapter 1) and ions and solvent also take time to reorganize within the IHP/OHP and diffuse layer. The process is not instantaneous since ions have to move, desolvate, solvate, and so on, and takes typically a few milliseconds. Hence, the total charge passed in response to the potential step will do so with an initial rapid increase and a less rapid decrease in current (Fig. 5.5, left). After this time, the current will be zero as no further changes occur and hence charge does not move.

In this section, we will assume that the redox active species present is in the reduced form and that we start from a potential more negative than the species $E_{½}$.

If instead of making a sudden step in potential, we change the potential of the electrode smoothly over time then the double layer will reorganize continuously in response. Instead of a transient flow of current for a short time, we will now see a steady current as we change the potential. If we stop and change the potential in the opposite direction then the double layer will reorganize again (but in the reverse way) and hence current will flow in the opposite direction. The ability of the solvent to counteract the polarization of the electrode is expressed as its double layer capacitance (C_{dl}) and the current that flows (i) is related to the rate of movement of charge (Q). If we think back to Chapter 1, we will remember that the charge built up in a capacitor is related to the voltage drop across it (i.e. $Q = C\,E$) and since current is the rate of passage of charge with time ($i = dQ/dt$) then '$i = C_{dl}\, dE/dt$'. Hence, the current is proportional to the rate of change in potential (referred to as the *scan rate* in V s^{-1}). If instead of plots of potential versus time and current versus time we plot current versus potential, then we will see a constant current with the sign (+/–) dependent only on the direction of the voltage ramp (Fig. 5.5).

Although this experiment can be represented by two plots; potential versus time (with a slope of $\frac{V}{t}$) and current vs time (with a slope of $\frac{i}{t}$), we can make use of a mathematical property, that is, $\left(\frac{di}{dt}\right)/\left(\frac{dE}{dt}\right)=\left(\frac{di}{dE}\right)$, and plot current versus potential ($\frac{i}{V}$) as shown in the linear sweep voltammogram.

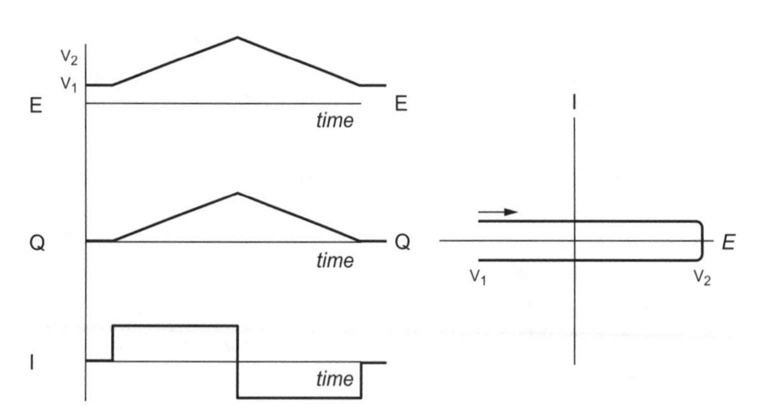

Figure 5.5 (left) Charge passed (Q) and current over time during a potential sweep from V1 to V2 and back to V1. (right) Resulting plot of current vs potential.

5.2.2.2 Current response to a potential step with a redox active species adsorbed to the surface

At any initial potential (V_1), the concentration of the oxidized and reduced forms of a species adsorbed to electrode (e.g. a thin polymer film or self-assembled monolayer) will conform to the Nernst equation. For simplicity, we assume that the initial potential (V_1) is negative compared to the equilibrium potential (i.e. $\eta < -500$ mV), and therefore the redox active species is in its reduced state.

The term equilibrium potential is sometimes used synonymously with 'formal potential' or $E^{o}_{1/2}$, but is not exactly the same. We will discuss this point later in the chapter, in Section 5.2.7.

When the potential is changed suddenly in a single step to another potential (V_2) that is more positive that V_1, then the ratio of oxidized and reduced forms of the species at the electrode will no longer correspond to that predicted by the Nernst equation; that is, the electrode will no longer be at equilibrium. Equilibrium is restored by a net transfer of electrons to or from the electrode to the species adsorbed, and a net Faradaic current flows as the system reaches the new equilibrium situation defined by the Nernst equation and V_2.

If the change in overpotential was not sufficient to make a significant change in the concentrations of oxidized and reduced species (i.e. V_2 is such that $\eta < -200$ mV), then essentially only non-Faradaic current due to the reorganization of the double layer will be observed (see Section 5.2.1.1). It is only when the new potential is in within around 100 mV (i.e. -100 mV $< \eta <$ 100 mV) of the equilibrium potential of the redox active species that a significant Faradaic current will flow. The flow of charge (current) will be transient and the current will return to zero once the double layer has reorganized and the equilibrium ratio of the oxidized and reduced forms of the redox active species is re-established.

5.2.2.3 **Current response to a linear potential sweep with a redox active species adsorbed to the surface**

If, instead of a sudden step, we change the potential of the WE slowly in a continuous manner (i.e. we apply a voltage 'ramp') then we will see a steady current due to the double layer (non-Faradaic current) charging, as discussed in Section 5.2.1.1, when the overpotential (η) is far from the equilibrium potential of the adsorbed redox active species. As the applied potential approaches the equilibrium potential the ratio of oxidized and reduced forms on the surface begins to adjust contributing to the measured current. From the Nernst equation we can see that the ratio of concentrations is dependent exponentially on the overpotential and hence there is an nonlinear increase in the amount of charge that needs to be passed to re-establish the equilibrium concentrations. Once we approach the equilibrium potential (as the ratio reaches 1:1) and, thereafter, a nonlinear decrease in current is observed as there is less and less change in concentrations needed (see Fig. 5.6). The shape of the plot of current versus potential is predictable provided that the rate of heterogeneous electron transfer between the electrode and surface confined species, and hence the continuous readjustment to conform to the Nernst equilibrium, is much faster than the rate of change of potential.

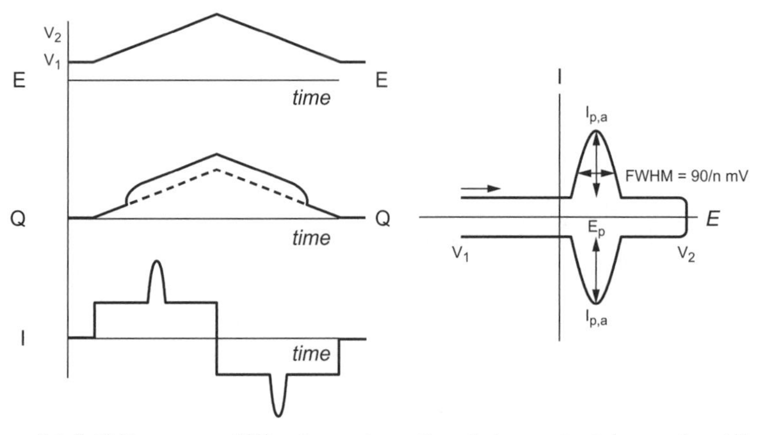

Figure 5.6 (left) Charge passed (Q) and current over time during a potential sweep from V1 to V2 and back to V1. (right) Resulting plot of current vs potential. The dashed line indicates charge passed in the absence of the redox active species adsorbed to the surface. Note the shape of the curves are exaggerated compared to actual CVs (see Box 5.1).

The shape of the voltammogram can be predicted using the Nernst equation. The current at any potential depends on the amount of charge passed per unit time. As the equilibrium potential is approached the charge that is passed will increase as the rate of change in concentration of the oxidized and reduced forms at the electrode/solution interface increases. The current at any potential is thus proportional to the rate at which the surface concentration Γ_o^* changes, the number of electrons transferred per mole of reactant and the area of the electrode (A),

$$I = nFA\frac{d\Gamma_{ox}}{dt} = -nFA\frac{d\Gamma_{red}}{dt}$$

We can use the Nernst equation to calculate Γ_{ox}

Algebraic rearrangement of the Nernst equation $E = E^{0'} - \frac{RT}{nF}\ln Q$ gives

$$\frac{-nF(E-E^{0'})}{RT} = \ln Q \text{ and since } Q = \frac{\Gamma_{red}}{\Gamma_{ox}} = \frac{\Gamma_{red}}{\Gamma_o^* - \Gamma_{red}} \text{ then}$$

$$exp\left\{-\frac{n\,F\,(E-E^{0'})}{RT}\right\} = \frac{\Gamma_{red}}{\Gamma_o^* - \Gamma_{red}}$$

A series of algebraic rearrangements on this equation allows us to express the concentration of the reduced form Γ_{red} as function of potential $(E - E^{0'})$ and total surface concentration (Γ_o^*) as follows:

$$(\Gamma_o^* - \Gamma_{red})exp\left\{-\frac{nF(E-E^{0'})}{RT}\right\} = \Gamma_{red}$$

$$\Gamma_o^* exp\left\{-\frac{nF(E-E^{0'})}{RT}\right\} - \Gamma_{red} exp\left\{-\frac{nF(E-E^{0'})}{RT}\right\} = \Gamma_{red}$$

$$\Gamma_o^* exp\left\{-\frac{nF(E-E^{0'})}{RT}\right\} = \Gamma_{red} exp\left\{-\frac{nF(E-E^{0'})}{RT}\right\} + \Gamma_{red}$$

$$\Gamma_o^* exp\left\{-\frac{nF(E-E^{0'})}{RT}\right\} = \Gamma_{red}\left(1 + exp\left\{-\frac{nF(E-E^{0'})}{RT}\right\}\right)$$

$$\frac{\Gamma_o^* exp\left\{-\frac{nF\,(E-E^{0'})}{RT}\right\}}{\left(1 + exp\left\{-\frac{nF\,(E-E^{0'})}{RT}\right\}\right)} = \Gamma_{red} \text{ and similarly } \frac{\Gamma_o^* \exp\left(\frac{nF(E-E^{0'})}{RT}\right)}{1 + \exp\left(\frac{nF(E-E^{0'})}{RT}\right)} = \Gamma_{ox}$$

The rate of change of Γ_{ox} with time is given by the first derivative with respect to time:

$$\frac{d\Gamma_{ox}}{dt} = \frac{d\left\{\dfrac{\Gamma_o^* \exp\left(\dfrac{nF(E-E^{0'})}{RT}\right)}{1+\exp\left(\dfrac{nF(E-E^{0'})}{RT}\right)}\right\}}{dt} = \Gamma_o^* \frac{d\left\{\dfrac{\exp\left(\dfrac{nF(E-E^{0'})}{RT}\right)}{1+\exp\left(\dfrac{nF(E-E^{0'})}{RT}\right)}\right\}}{dt}$$

Differentiation by parts (see Exercise 5.2 at the end of this chapter) and taking into account that the scan rate $v = dE/dt$:

$$I = nFA\frac{d\Gamma_{ox}}{dt} = \frac{n^2F^2vA\Gamma_o^*}{RT}\frac{\exp\left(\dfrac{nF(E-E^{0'})}{RT}\right)}{\left(1+\exp\left(\dfrac{nF(E-E^{0'})}{RT}\right)\right)^2}$$

The characteristic features of the cyclic voltammogram recorded when a redox active species is present on the electrode surface are the maximum (peak) currents $I_{p,a}$ and $I_{p,c}$ (Fig. 5.1) and the full width at half maximum of the anodic and cathodic waves (3.53 RT/nF or 90/n mV where n = number of electrons transferred in a reversible redox process). If the redox process is chemically and electrochemically reversible then $I_{p,a} = I_{p,c}$ and $E_{p,a} = E_{p,c}$. The peak currents are obtained when $\eta = 0$ (i.e. the exponential terms are all equal to 1).

$$I_p = \frac{n^2F^2}{4RT}vA\Gamma_o^*$$

which is:

$$I_p = (9.39 \times 10^5)n^2vA\Gamma_o^*$$

at 25°C, where Γ_o^* is the total surface concentration in mol cm^{-2}, A is the electrochemical surface area, v is the scan rate in V s^{-1} and n is the number of electrons involved in the redox process. The peak current is linearly dependent on the scan rate, v, which is a characteristic that is useful in confirming the redox process is due to a surface confined species (see Box 5.1).

Furthermore, the area under the redox wave due to the Faradaic current is the total charge passed and hence:

$$\frac{\textit{Area under curve}}{v} = nFA\Gamma_o^*$$

This relation allows us to use cyclic voltammetry to determine the surface density (typically 10^{-10} mol cm^{-2} for a monolayer of molecules) of the redox active species; for example, as shown in Box 5.1.

$I = \frac{dQ}{dt}$ and hence $\int I dt = Q$ and since time is directly proportional to potential when $\frac{dE}{dt}$ is constant, $\int I dE = c\ Q$ where c is a constant.

Box 5.1 Cyclic voltammetry of redox polymer modified electrodes

The linear sweep and, as we will meet later in the chapter, cyclic voltammetry of a redox active material or compound on a surface is often easily recognized from the shape of the cyclic voltammogram. For example, when a polymer, which is made up of a chain of redox active subunits called 'double spiropyrans', is coated onto a glassy carbon electrode, the subunits undergo two one-electron oxidations at 0.65 and 0.78 V versus SCE, respectively, manifested in two redox waves in the cyclic voltammogram (Fig. 5.7). Removal of one electron is energetically easier than removal of the second electron, which is reflected in the potential of each process (remember that potential is related directly to Gibbs energy change).

The original publication can be found using the DOI: 10.1021/jacs.5b11604

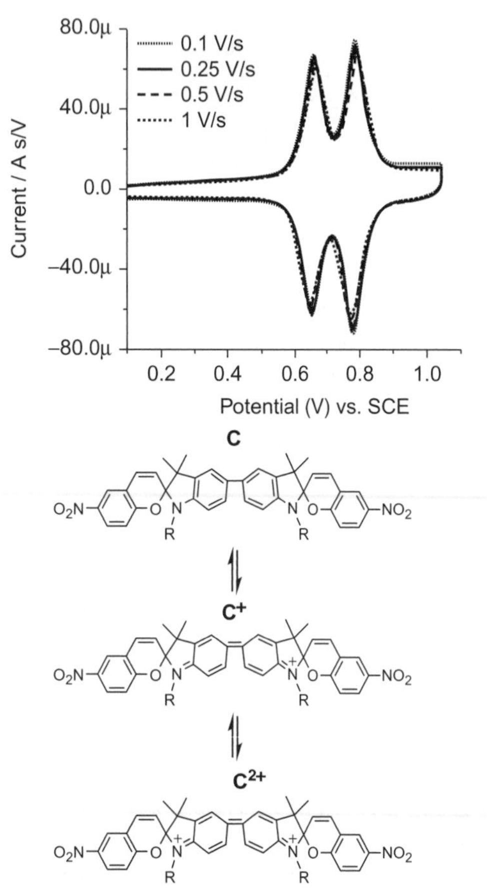

Figure 5.7 Cyclic voltammogram of a redox active polymer (the repeat unit is shown below) coated glassy carbon electrode (3 mm diameter) showing close to ideal behaviour; the separation between $E_{p,a}$ and $E_{p,c}$ for each redox process is close to zero and the full width at half maximum is in each case close to 90 mV. The first redox wave is due to the C/C^+ redox couple and the second is due to the C^+/C^{2+} redox couple. The second oxidation requires a more positive electrode potential than the first. The scan rate is varied between 0.1 and 1.0 V s^{-1}, however, if we divide the current in A by the scan rate (V s^{-1}) then we see that the voltammograms are almost identical.

The surface coverage (Γ) can be calculated by integrating the area under either of the redox waves and shows that the film is only a few monolayer equivalents thick. In this case the potentials ($E_{p,a}$ and $E_{p,c}$) at which the maximum current ($I_{p,a}$ and $I_{p,c}$) is observed are identical regardless of scan direction and both of the redox waves have close to a 90 mV full width at half the maximum current (FWHM). Since there is a finite amount of material on the electrode, there is also a limit to the total Faradaic charge that can pass when the potential is switched from negative to positive overpotentials. Hence the total amount of charge passed (Q) is not dependent on scan rate (the rate of change of potential) but the current is linearly dependent on scan rate. This aspect means that a further characteristic of a surface confined redox process is that the cyclic voltammograms recorded over a range of scan rates will be nearly identical when the current axis is normalized to the scan rate.

5.2.2.4 **Current response to a potential step for a species where both oxidized and reduced forms are soluble**

We now consider a more complex situation where the redox active species is present in solution (i.e. soluble) in both its oxidized and reduced states. In our analysis of this situation, we have to keep several key points in mind:

- The ratio of reduced and oxidized forms at the electrode/solution interface will be determined by the potential the electrode is held at and can be calculated using the Nernst equation:

$$\exp\left(-\frac{nF\left(E-E_{\frac{1}{2}}\right)}{RT}\right)=\exp\left(-\frac{nF(\eta)}{RT}\right)=\frac{[Red]}{[Ox]}$$

- The concentration of oxidized and reduced forms in the bulk solution (C_o^*) is not necessarily equal to that at the electrode and if it is not then there will be a concentration gradient between the electrode and the edge of the depletion layer.
- Although the change in concentrations at the electrode, in response to a change in potential, is rapid, the development of an essentially steady state concentration gradient between the electrode and the edge of the stagnant layer takes a finite length of time. The growth of the Nernst diffusion (depletion) layer occurs on time scales that are similar to typical voltage ramping rates (scan rates, e.g. 0.1 V s^{-1}).

In our analysis of this situation, we will begin by working under conditions where the rate of change in potential is sufficiently low such that the build-up of the Nernst diffusion layer is essentially complete at all times.

If we use a micro-electrode (with a diameter of 10–50 μm) then radial diffusion will dominate mass transport. The electrode dimensions are much less than the typical thickness of the diffusion layer and species formed at the electrode diffuse away from it rapidly, while species consumed are replaced rapidly from the bulk solution. The shape of the voltammogram will be the same as that obtained with a cyclic voltammogram at a mini or macro-electrode at low scan rates (e.g. 1 mV s^{-1}).

5.2.3 **Linear sweep voltammetry at low scan rates**

As discussed in previous chapters, when an electrode is at equilibrium at a particular potential (V_1) then the concentration of species at the electrode solution interface satisfies the Nernst equation. If the concentrations of oxidized and reduced forms of a redox active species at the electrode are equal to their respective concentrations in the bulk solution then there will be no net current flow; this potential is referred to as the OCP.

The measurement of the OCP is a useful method to determine the ratio of oxidized and reduced forms of a species in solution. Note that this potential is not 'zero volts'! It is good practice to set the initial potential in a voltammogram to the OCP and not to begin at the '0.0 V'.

Let us begin the discussion of linear sweep voltammetry by considering a situation where only the reduced form of a redox active species is present in solution. In this situation [Red] >> [Ox], and from the Nernst equation we can see that V_1 (the OCP) will be at negative overpotentials.

If we change the applied potential to a less negative overpotential, but are still far from the equilibrium potential of the redox couple, then the current will be due almost completely to double layer charging (i_c) since the concentrations of reduced and oxidized forms at the electrode will differ little, if any, from that in the bulk. The concentration gradient established over the diffusion layer will, therefore, be negligible.

If we change the potential in a single step such that the electrode is has now a positive overpotential (V_2) then the concentrations of oxidized and reduced forms at the electrode will no longer satisfy the Nernst equation. A net current (a Faradaic current) will flow to bring the concentration of the oxidized and reduced forms at the electrode to that dictated by the electrode potential (V_2) through the Nernst equation. In contrast to a surface confined species though (see Section 5.2.2), in this new situation a concentration gradient ($\frac{d[Red]}{dx}$ where x is the distance from the electrode) is established between the electrode and the bulk since the concentrations of the species at the electrode will be different to their concentrations in the bulk. There will be a net diffusion of reduced species to the electrode and the oxidized species away from the electrode to 'fight' this difference in concentration. Hence, a current must flow to maintain the equilibrium concentrations at the electrode solution interface dictated by the Nernst equation and the applied potential.

Now let us take a detailed look at what happens over time following such a potential step. Initially, the concentration changes only at the electrode within the first few nanometres and hence the concentration gradient is steep. Over time the concentration gradient will decrease as the Nernst diffusion layer is established and eventually the edge of the stagnant layer will be reached—the Nernst diffusion layer is then said to be **fully developed**, as shown in Fig. 5.8.

The net rate of electron transfer (and therefore current) will depend on the net flux of the reduced species diffusing towards the electrode and the oxidized species diffusing away. If we assume the diffusion coefficients for both oxidized and reduced forms are equal then the rate of electron transfer (i.e. current) is proportional to the concentration gradient at the electrode ($x = 0$); as well as the number of electrons transferred (n), Faraday constant (F, C mol^{-1}), the electrode area (A, cm^2) and the net flux of the reduced form to the electrode ($D_A \left(\frac{d[Red]}{dx}\right)_{x=0}$, cm^2 s^{-1} M^{-1} cm^{-1}) by:

It is handy to consider dimensional analysis in this equation just to check we have covered all factors contributing to the current correctly. I is $A = C\,s^{-1} = C\,mol^{-1} * cm^2 * cm^2\,s^{-1} * mol\,dm^3\,cm^{-1} = C\,s^{-1} * mol^{-1} * mol * cm^2 * cm^2\,(1000)\,cm^3\,cm^{-1}$.

$$I = n\,F\,A\,D_A \left(\frac{d[Red]}{dx}\right)_{x=0}$$

We will justify this equation next.

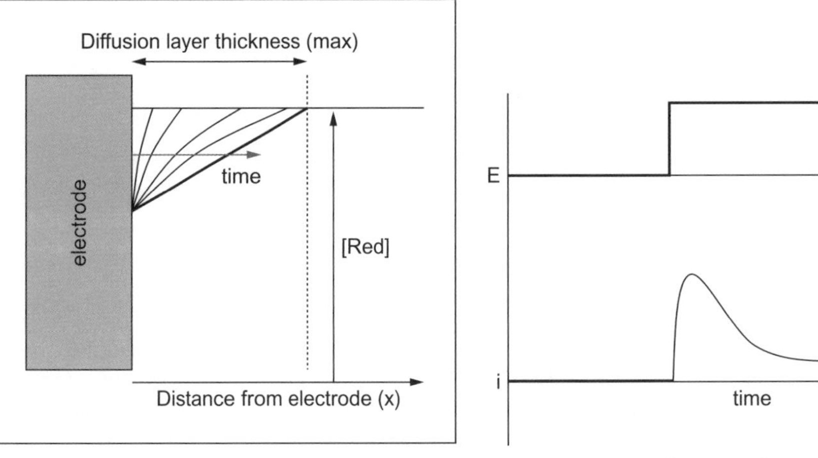

Figure 5.8 (left) The development of the concentration gradient at an electrode follow a step from a large negative overpotential to an overpotential close to the half wave potential. Note that the vertical axis '[Red]' represents the concentration of the reduced form from 0 to the bulk concentration (C_0*). (right) The step in potential results initially in a very steep increase in current followed by a more gradual decrease in current as the Nernst diffusion layer develops to the thickness of the stagnant layer.

It is apparent that immediately after the potential is changed the concentrations at the electrode solution interface (IHP/OHP/diffuse layer) are changed to the equilibrium ratio but close to the electrode, the concentrations are still the same as in the bulk solution. Therefore, the concentration gradient $\frac{d[Red]}{dx}$ is steep and the current is high. Within a short time however the concentration changes near the electrode and the concentration of the reduced form is decreased further and further into the depletion layer develops. The consequence of this change is that the gradient $\left(\frac{d[Red]}{dx}\right)$, and therefore the current, continues to decrease over time until the concentration gradient has reached the edge of the stagnant layer or when convection is negligible the rate of decrease in the slope will be negligible on the time scale of the experiment.

The current increases rapidly (depending on the RC time constant) and then decreases with a $1/t^{1/2}$ dependence as the depletion layer develops until it reaches the depth of the stagnant layer (δ). This time dependent change after the current reaches a maximum is described by the Cottrell equation.

Although the thickness of the stagnant layer might not be well defined, for example when convection is not forced, the rate of decrease in current will eventually be negligible and therefore a fixed depth can be assumed as a good approximation.

5.2.4 **The Cottrell equation**

Although we will not derive the Cottrell equation here, it is informative to consider its origin and the rationale behind the equation. For a large area planar electrode, the rate of change of concentration $\left(\frac{dC_{x,t}}{dt}\right)$ is dependent on the diffusion coefficient of the species and can be predicted at any distance from the electrode and time by consideration of Fick's laws of diffusion in one dimension:

$$\frac{dC_{x,t}}{dt} = D\,\frac{d^2C_{x,t}}{dx^2}$$

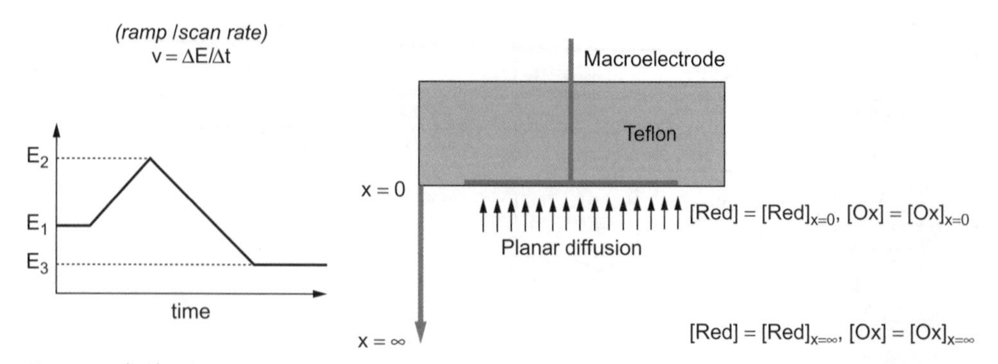

Figure 5.9 (left) Voltage ramp with change in direction followed during a cyclic voltammogram and (left) boundary conditions for concentration of oxidized and reduced forms of a species at the electrode ($x = 0$) and at an infinite distance from the electrode (which is in reality the boundary between the stagnant layer and the bulk solution).

From a molecule's point of view (if a molecule would have a point of view of course!), if the diameter of an electrode is large (e.g. 3 mm) compared to the distance it can diffuse along the plane of the electrode during the voltammogram, then we can make a simplification that only diffusion orthogonal to the plan of the electrode is important since diffusion of all molecules in the plane will not make a net difference to concentration.

where x (cm) is the distance from the electrode and D is the diffusion coefficient (cm^2 s^{-1}). Taken together with the Nernst equation, the current can be predicted.

At any electrode potential, the equilibrium concentrations of the redox active species at the electrode can be calculated by the Nernst equation:

$$\left(-\frac{nF\left(E-E_f^o\right)}{RT}\right)=-\frac{nF(\eta)}{RT}=\ln\frac{[Red]_{x=0}}{[Ox]_{x=0}}$$

where $x = 0$ means at the electrode (i.e. the Helmholtz plane) and the net flux of 'Red' is equal but opposite to the flux of 'Ox' at the electrode.

$$\frac{d[Red]_{x=0\ \ t>0}}{dt}=-\frac{d[Ox]_{x=0\ \ t>0}}{dt}$$

The concentration of the redox active species at the electrode changes almost instantaneously in response to a change in electrode potential. However, the concentration at any distance from the electrode is time dependent (i.e. the Nernst diffusion layer thickens over time).

In the discussion here we will consider a situation where the redox active species is present in the bulk in only one oxidation state and refer to the concentration in the bulk solution as C_o^* and the concentration at a distance x from the electrode at a time t as $C_{x,t}$.

The current will depend on the area of the electrode (A) and the diffusion coefficient (D). As mentioned before, ultimately (e.g. as x becomes large, i.e. the thickness of the stagnant layer, see Chapter 3), the concentration will reach the concentration in the bulk solution, that is $\lim_{x\to\infty} C_{x,t} = C_o^*$. Hence, the concentration in regions beyond the stagnant layer are not affected by changes in electrode potential since their concentration is maintained by mass transport by convection and not diffusion.

Within the Nernst diffusion layer, that is the boundary conditions of $C_{x=\infty,t} = C_o^*$ and $C_{x=0,t} = 0$ at a large overpotential (positive where the reduced form is present in solution and vice versa) at $t > 0$ (Fig. 5.9), the concentration at any point can be calculated by considering δ is the thickness of the layer between the electrode and the plane where the concentration is C_o^*. The current is dependent on the net speed with which the species in solution diffuse to the electrode. The net flux of molecules (J_x) towards the electrode is related to the current (density) and hence at the electrode:

$$-J_{x=0,t} = D\frac{dC_{x=0,t}}{dx} = i_t/nFA$$

where $-J_{x=0,t}$ has units of mol cm^{-2} s^{-1}; D is in cm^2 s^{-1}, $dC_{x=0,t}/dx$ is mol dm^{-3} cm^{-1}; and *it* is C s^{-1}, *F is* C mol^{-1}, and *A* is cm^2 and hence a rearrangement gives the equation we met previously:

$$i_t = nFA\,D\frac{dC_{x=0,t}}{dx}$$

From this we can consider the current at any time, which leads to the Cottrell equation:

$$i_d(t) = n\,FA\,D^{½}\,C_o/\pi^{½}\,t^{½}$$

Examination of the Cottrell equation would lead to the conclusion that eventually (as *t* becomes very long) the current should approach zero. However, the diffusion layer does not thicken indefinitely (even when convection is not forced). Once the concentration gradient has relaxed to the thickness of the stagnant layer (δ_{max}), the concentration gradient does not decrease further. Finally, the current will remain constant at a level dictated by the rate of net diffusion between the bulk solution and the electrode (diffusion limited current) and is therefore no longer time dependent.

As a last remark, current is time dependent (i vs. $1/t^{½}$) and therefore we can determine diffusion coefficients by chronoamperometry; that is, measuring the current response over time following a step change in electrode potential.

5.2.5 **Mass transport limited current**

Although in Chapter 4, we discussed the net rate of oxidation/reduction in terms of overpotential, its effect on the energy of activation, and thereby on the rates of heterogeneous electron transfer, for electrochemically reversible redox processes diffusion often limits the maximum current (net rate of electron transfer) that is observed in a cyclic voltammogram. The rate at which a species diffuses from the bulk to the electrode determines the net flux of the species to the electrode. Fick's second law of diffusion allows us to calculation the concentration gradient established:

$$\frac{dC_{x,t}}{dt} = D\frac{d^2C_{x,t}}{dx^2}$$

For simplicity, we will assume that only the reduced form of the species is in solution and that the diffusion layer is established fully (and equal to the stagnant layer thickness, δ_{max}), that is $C_{x,t} = [Red]_x$ (where x is the distance from the electrode). Under these conditions, the concentration gradient between the bulk solution and the electrode is time independent and equal to:

$$\frac{d[Red]}{dx} = \frac{\{[Red]_o - [Red]_{x=0}\}}{\delta_{max}}$$

(where $[Red]_o$ is the bulk concentration) and since the rate of oxidation (manifested as net current flow) is:

$$I = n\,F\,A\,D_{Red}\left(\frac{d[Red]}{dx}\right)_{x=0}$$

Then:

$$I = \frac{nF\ A\ D_{Red}}{\delta_{max}}\{[Red]_o - [Red]_{x=0}\}$$

Where the terms $\frac{nF\ A\ D_{Red}}{\delta}$ are constants and can be combined as a 'rate constant' for oxidation and hence when the overpotential is large and $[Red]$ at the electrode is essentially zero then the diffusion limited current is $I_{lim} = nFA\left(\frac{D_{Red}}{\delta_{max}}\right)[Red]_o$

It is important to note that under these conditions $\left(\frac{D_{Red}}{\delta_{max}}\right)$ has the same units as the rate constant for heterogeneous electron transfer discussed in Section 4.3, which allows for direct comparison of their magnitude and hence whether the current is limited by mass transport or by electron transfer kinetics. Furthermore, it should be apparent that current will increase if we reduce the thickness of the stagnant layer (δ_{max}) by, for example, by a wall jet or rotating disc electrode.

In this section we make the assumption that convection is forced at the electrode and the thickness of the stagnant layer is well-defined. The general conclusions hold to a large extent for situations where the stagnant layer is the entire solution (i.e. when convection is minor or absent).

5.2.6 **Current response to linear potential sweeps where both oxidized and reduced forms of a redox active species are soluble**

We now consider an experiment where we change the voltage continuously with time instead of in discrete steps; that is, where dE/dt is a constant. As discussed previously, if the solution does not contain species that are redox active in the potential range or 'window' (i.e. between the initial and final potential of the ramp) then the current measured will be essentially constant and due to the continuous reorganization of the Helmholtz planes (i.e. capacitance). The magnitude of the current in this case can be calculated based on the interfacial capacitance, $i = C_{dl}\ dE/dt$.

In the presence of a redox active species, we now have to consider a situation where the rate of change of potential (the scan rate; $\mathbf{v} = dE/dt$) is sufficiently slow, such that at all times:

- The relative concentrations of the oxidized and reduced forms at the electrode surface conform to that dictated by the Nernst equation.
- The maximum thickness of the diffusion layer is established (δ) essentially instantaneously.

If the initial potential is far from the half-wave potential ($E_{½}^{o}$) of the redox active species then the current will be constant over time. As the potential approaches

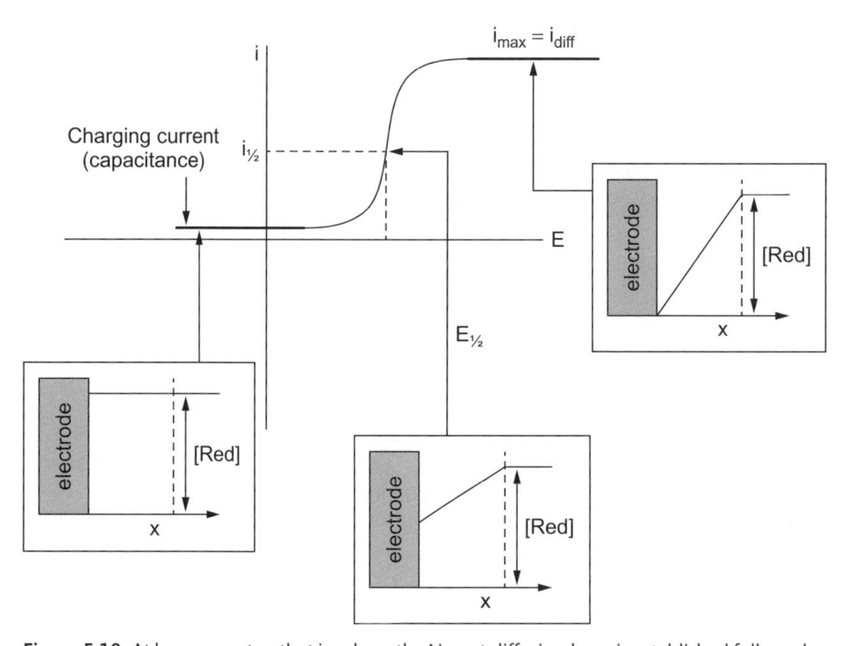

Figure 5.10 At low scan rates, that is, where the Nernst diffusion layer is established fully and the concentration gradients are at a minimum at all times, the current is solely dependent on the concentration gradient and hence overpotential. Note that the vertical axis '[Red]' represents the concentration of the reduced form from 0 to the bulk concentration (C_o*).

$E_{½}^{o}$, the equilibrium position between oxidized and reduced forms will change significantly and the current will increase as the concentration gradient between the electrode and the bulk solution increases. As the potential goes beyond $E_{½}^{o}$ the rate of increase in current lessens and eventually [*Red*] at the electrode is effectively zero. At overpotentials >200 mV, the concentration gradient is unaffected by the electrode potential and the current is constant. The result is a sigmoidal shaped curve, the shape of which is independent of scan direction with the maximum current is diffusion limited. The potential at which half the maximum current is reached corresponds to $E_{½}^{o}$.

The half-wave potential ($E_{½}^{o}$) is the potential at which [Red]=[Ox] at the electrode and is close to the formal potential (E_f^{o}), see Section 5.2.7 for detailed discussion.

Strictly speaking, this is only correct if the diffusion coefficients of the oxidized and reduced forms are the same or similar.

The maximum current is limited by the rate of diffusion of the reduced form of the species to the electrode and hence the concentration gradient that is established between the electrode and the edge of the Nernst diffusion layer. For micro-electrodes and macro-electrodes (at low scan rates) the diffusion layer reaches its maximum extent essentially instantaneously (compared with the time scale of the change in potential, Fig. 5.10). The charging current (electrode capacitance, which depends on the rate of change of electrode potential) is negligible usually under these conditions and the maximum current is limited by diffusion.

If the overpotential is sufficient to decrease the equilibrium concentration of the reduced form at the electrode to zero, that is $[Red]_{x=0} = 0$, then the diffusion limited current will be $I_{lim} = k_{red}\,[Red]_o$ and will depend solely on the concentration of the reduced form in the bulk solution as discussed previously. The diffusion limited current and its direct relation with concentration is the basis for several electrochemical sensors as discussed in Box 5.2.

Typically, the OCP is chosen as E_1 so that chemistry does not occur at the electrode (i.e. the concentrations of all species are equal to their bulk concentrations) before the experiment begins. E_2 is the first switching potential and E_3 is the final potential. The maximum scan rate that can be used is limited by the RC time constant and the effect of iR_{drop} (note that higher scan rates lead to higher currents and an increased contribution from uncompensated resistance).

Box 5.2 Electrochemical sensors

The electrochemical sensors based on measuring current flow are a major application of dynamic electrochemistry, as much as the pH meter is for potentiostatic electrochemistry. The most important, in terms of impact, are without doubt the Clark oxygen electrode and the glucose sensor (which unfortunately is ubiquitous due to the global challenge presented by diabetes).

The reduction of molecular oxygen to superoxide proceeds readily at –0.8 V versus Ag/AgCl, and its concentration in solution or the gas phase can be determined from the diffusion limited current by placing a membrane permeable to oxygen only over a platinum or gold electrode (e.g. the Clark electrode). Clark and Lyons developed a glucose sensor in the 1960s using the oxygen electrode, with a dialysis membrane containing the enzyme glucose oxidase. The enzyme consumes oxygen as it converts glucose and the decrease in oxygen concentration is reflected in a decrease in current at the Clark oxygen electrode. The addition of a second electrode without the enzyme in the membrane allows for correction for current due to other redox active species present.

dialysis membrane containing glucose oxidase

Clark O_2 electrode

glucose + O_2

electrode

gluconolactone + H_2O_2

H_2O + ½O_2

Although glucose is oxidized in the process the direct electrochemical oxidation of glucose is not feasible due to an unfeasibly high overpotential and hence the use of an enzyme is critical. Heller and his team developed an electrochemical based glucose sensor that did not rely on oxygen reduction but instead used enzymes to extract electrons from glucose. The key challenge for such an approach was to allow the enzyme to transfer electrons to the electrode, which due to the size of the enzymes and the 'buried' active site meant that the low rates of heterogeneous electron transfer needed to be overcome. This was achieved by using a redox mediator–a small molecule that could take the electrons from the enzyme and pass it to the electrode (in the same way a bucket brigade works to bring water to a fire).

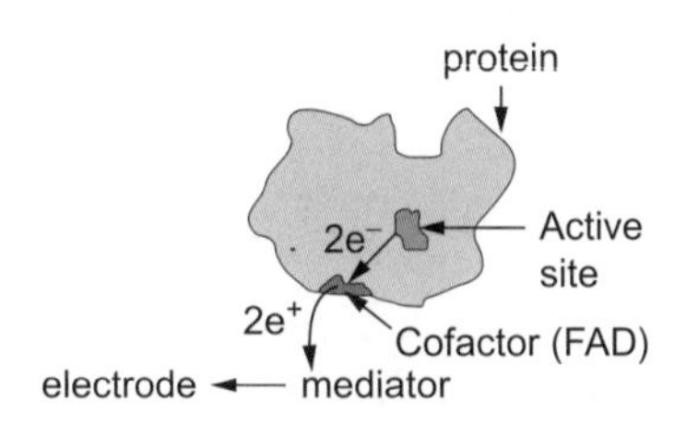

The mediator most commonly employed is a water soluble derivative of ferrocene, ferrocene monocarboxylic acid (FMCA), which is chemically stable in both oxidation states, has large k^{o} and D ($3 \times 10^{-6} cm^2\ s^{-1}$). The reactions are:

$$\text{Glucose} + \text{GOD}_{ox} \xrightarrow{k_1} \text{gluconolactone} + \text{GOD}_{red}$$

$$2\ \text{FMCA}^{+} + \text{GOD}_{red} \xrightarrow{k_{cat}} 2\ \text{FMCA} + 2\text{H}^{+} + \text{GOD}_{ox}$$

$$2\ \text{FMCA} \overset{E_F^{o},\, k^{o}, \alpha}{\leftrightarrow} 2\ \text{FMCA}^{+} + 2\ e^{-}$$

where k_1 and k_{cat} are homogenous rate constants.

It should be noted that the reaction between $FMCA^+$ and the GOD_{red} is in fact a reaction with the reduced form of the cofactor, that is $FADH_2$, which is bound to the enzyme. Importantly the $E_{½}$ of the FMCA/$FMCA^+$ redox couple is positive of the potential of the $FADH_2/FAD^+$ redox couple (i.e. the driving force for oxidation of $FADH_2$ by $FMCA^+$ is large).

If there is no glucose present, the enzyme does not reduce the $FADH_2$ cofactor and hence a normal reversible cyclic voltammogram of FMCA is observed with the peak currents defined by the Randles-Sevĉik equation (see section 5.2.7). However, if the concentration of the enzyme and hence $FADH_2$ is in excess of the mediator $FMCA^+$, then in the presence of glucose, any $FMCA^+$ produced at the electrode will be reduced rapidly by the enzyme bound $FADH_2$ within the Nernst diffusion layer preventing the establishment of a fully depleted Nernst diffusion layer. The consequence of this is the observation of a so-called catalytic current, which is limited only by the concentration of glucose.

The current measured when glucose is present depends on the rate at which FMCA diffuses to the electrode and the rate at which is produced with in the diffusion later by reduction of $FMCA^+$ by the enzyme. The higher the glucose concentration then the higher the rate of reduction of $FMCA^+$. This relation can be expressed by the equation:

$$I = \frac{nFAC\sqrt{D\,k_{cat}^{'}}}{1 + \exp\left[\frac{nF\left(E - E_{\frac{1}{2}}\right)}{RT}\right]}$$ where C is the concentration of ferrocene carboxylic acid

When $I = I_{lim}$ (the limiting current), that is, when the overpotential is large, then:

$$\exp\left[\frac{nF\left(E - E_{\frac{1}{2}}\right)}{RT}\right] \approx 0$$

Hence $I = nFAC\sqrt{D\,k_{cat}^{'}}$, so $I \propto \sqrt{k_{cat}^{'}}$, and is independent of scan rate. Therefore, as long as the potential of the WE is held at a sufficient overpotential for the oxidation of FMCA and $k_1 << k_{cat}$ then the (catalytic) current then will depend only on the concentration of glucose (Fig. 5.11).

Modern glucose sensors are comprised of a small disposable electrode set (often with screen printed electrodes) on which a drop of blood is placed and the current is read out by a small portable electronic reader. Since the blood contains a large number of redox active components the current at two WE is read out simultaneously with only one of the electrodes containing the enzyme.

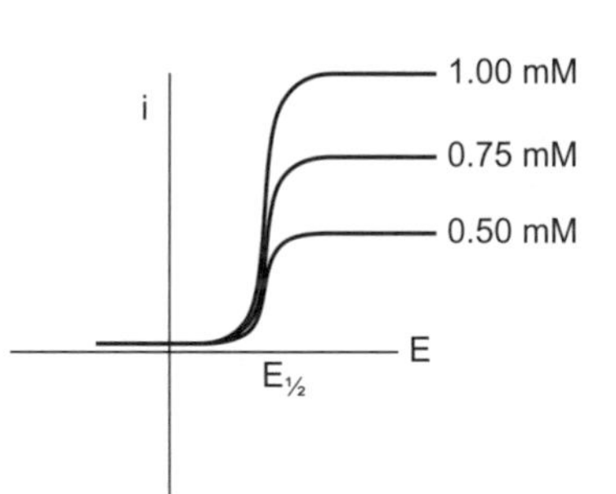

Figure 5.11 Effect of an increase in concentration of an oxidizable species ([Red]) on the limiting current observed at positive overpotentials.

5.2.7 **Linear sweep voltammetry at a macro-electrode or at high scan rates**

In contrast to voltammograms at low scan rates where the Nernst diffusion layer is established fully at all times, with macro-electrodes the scan rates typically used (0.1–1 V s^{-1}) are sufficiently high that the Nernst diffusion layer does not establish itself fully until large overpotentials are reached (Fig. 5.12).

During a linear sweep voltammogram the overpotential increases linearly, which results in a shift of the equilibrium position of the reaction $Ox + e^- \rightarrow Red$.

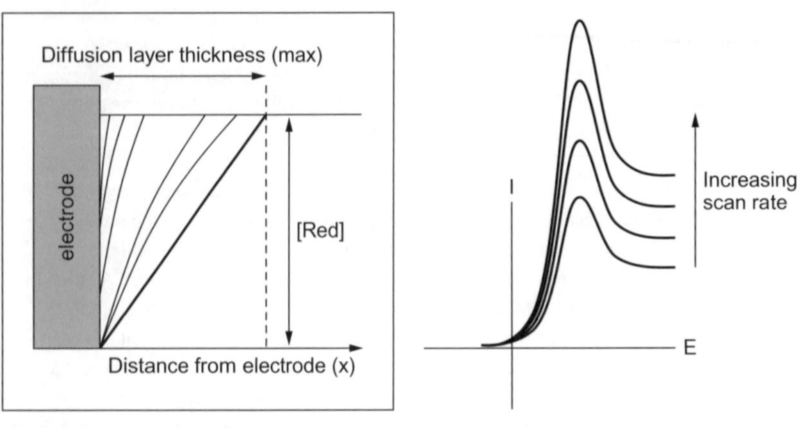

Figure 5.12 Effect of high potential sweep rates on the establishment of the concentration gradient during a linear sweep voltammogram. The change in potential and hence equilibrium position at the electrode is more rapid that the rate at which the Nernst diffusion layer is established. Note that the vertical axis '[Red]' represents the concentration of the reduced form from 0 to the bulk concentration (C_o^*). The corresponding linear sweep voltammograms at increasing sweep rate is shown on the right.

The name given to an electrode, micro, mini, and macro, refers casually to electrode area, with radii of, for example, 10 μm, 0.5 mm, and 3.0 mm, respectively. However, in making this classification it is better to refer to the equations required to describe the current flowing at the electrode; that is, radial diffusion for micro and planar diffusion for macro-electrodes.

For reversible redox processes the concentrations at the electrode change essentially instantaneously in response to the change in potential, resulting in a sudden increase in the concentration gradients (e.g. $d[\mathrm{Red}]/dx$) at the electrode ($x = 0$) and hence an increase in current. However, a decrease in the concentration of the reduced form 'Red' occurs rapidly also, which is manifested in the growth of the depletion layer towards the edge of the stagnant layer; a process that reduces the concentration gradient. It is the net result of this pair of processes that determines the current measured and while the time needed for the diffusion layer to be depleted and the slope of the concentration gradient to reach a minimum is fixed, the rate at which the overpotential is changed (scan rate) can be varied.

At low scan rates the Nernst diffusion layer is essentially completely established to the edge of the stagnant layer at all times and the concentration gradient increases steadily before reaching a maximum at around η = 100 mV. However, if the rate of change of potential is greater than the rate at which the Nernst diffusion layer is developed, and hence the concentration gradient, then the current will increase rapidly until the overpotential is such that the concentration of the species present in the bulk solution is zero at the electrode. Thereafter, the gradient will not increase further regardless of how much the overpotential is increased but instead the diffusion layer will develop (thicken) further and the concentration gradient decrease until the edge of the stagnant layer is reached.

In mathematical terms, if $[\mathrm{Red}]_{x=0} \rightarrow 0$, that is, the overpotential is increased, then $d[\mathrm{Red}]/dx$ increases, however, this is counteracted by an increase in the thickness of the diffusion layer (to δ) that decreases $d[\mathrm{Red}]/dx$. When $[\mathrm{Red}]_{x=0} = 0$ then $d[\mathrm{Red}]/dx$ decreases as the diffusion layer reaches its maximum thickness and then becomes constant thereafter (diffusion limited current).

Although the current at any potential can be predicted we typically focus on the peak currents obtained when scanning towards positive ($I_{p,a}$) and negative ($I_{p,c}$) potentials as these currents can be related to the concentration, electrode area and diffusion coefficient using the Randles-Sevĉik equation:

$$i_p = 0.4463\,(F^3/RT)^{½}\, n^{3/2}\, A\, D_o^{½}\, C^*\, v^{½}$$

where i_p is the peak current in amps (A), Faraday's constant in C mol^{-1}, A is area in cm^2, D_o is the diffusion coefficient in cm^2 s^{-1}, C^* is the bulk concentration and v is the scan rate in V s^{-1}.

At 25°C (for an electrochemically and chemically reversible process), the equation is:

$$I_{p,a} = (2.69 \times 10^5)\, n^{3/2}\, A\, D^{½}\, v^{½}\, C^*$$

It is apparent that $I \propto \sqrt{v}$, which is due to the insufficient time available to establish the Nernst diffusion layer before the overpotential has changed.

Furthermore, we should also remember that the charging current is:

$$I_c = A\, C_{dl}\, v$$

where A is the electrode area, C_{dl} is the double layer capacitance and v the scan rate in V s^{-1} and hence the ratio of the Faradaic and non-Faradaic current (colloquially called the background/baseline current) at $E_{p,a}$ (or $E_{p,c}$ for a sweep to negative potentials) is:

$$I_p/I_c = \{(2.69 \times 10^5)\, n^{3/2}\, D_A^{½}\, [A]_o\}/C_{dl}\, v^{½}$$

Hence, the ratio is inversely dependent on scan rate and at high scan rates, surface confined processes such as capacitance dominate a sweep voltammogram.

A key challenge to the interpretation of linear sweep (and cyclic, see below) voltammetric data lies in understanding why a voltammogram has a particular shape in various situations. In this section, we will consider the limiting situations where both the oxidized and reduced forms of a redox active species are free to diffuse to and from the electrode, and with the assumption that the redox chemistry:

$$\text{Ox} + e^- \rightarrow \text{Red}$$

is chemically and electrochemically reversible. Furthermore, at all times the ratio of oxidized 'Ox' and reduced 'Red' forms of the redox active species at the electrode will be determined by the overpotential (η) via the Nernst equation, that is, if the over potential is large and negative (> –200 mV) then the exponential term becomes very large and [Ox] << [Red].

Let's consider that the redox active species in solution is in the reduced state and hence the bulk concentrations are: [Ox] << [Red]. Hence, as the voltage is ramped from an initial negative potential towards positive potentials the ratio [Red]/[Ox] at the electrode does not initially change significantly and the Faradaic current will be negligible. The only current observed will be due to double layer capacitance. However, as the electrode potential approaches the half-wave potential (i.e. η approaches zero) then the ratio [Red]/[Ox] at the electrode will

Note that if $D_{red} = D_{Ox}$ then the half-wave potential $E_{½} = E_F^o$, whereas if $D_{red} \neq D_{Ox}$ then $E_{½} = E_F^o + (RT/nF)\ln(D_A/D_B)^{½}$, however, typically $E_{½}$ is within 10 mV of E_F^o. The diffusion coefficient can be quite different, for example, when a neutral species is oxidized since the cationic species formed is ion paired and more heavily solvated, for example, O_2/O_2^-.

$$\exp\left(\frac{-nF\eta}{RT}\right) = \frac{[red]}{[ox]}$$

be forced to change to 1:1, which is only achieved by electron transfer from Red to the electrode.

At this stage the current can be predicted (ignoring capacitance) by considering linear diffusion as the only source of mass transport between the bulk solution and the electrode and hence Fick's second law of diffusion:

$$d[\mathrm{Ox}]/dt = D_{\mathrm{ox}}\, d^2[\mathrm{Ox}]/dx^2$$

$$d[\mathrm{Red}]/dt = D_{\mathrm{Red}}\, d^2[\mathrm{Red}]/dx^2$$

Since only Red is present in the bulk solution ($[\mathrm{Red}]_o$) then at time t, and distance x from the electrode:

$$t = 0\ x > 0 [\mathrm{Red}] = [\mathrm{Red}]_o\ [\mathrm{Ox}] = 0$$

$$t > 0\ x \to \infty\ [\mathrm{Red}] = [\mathrm{Red}]_o [\mathrm{Ox}] = 0$$

$$t > 0\ x = 0\ D_{red}\,(d^2[\mathrm{Red}]/dx^2)_{x=0} = D_{ox}(d^2[\mathrm{Ox}]/dx^2)_{x=0}$$

that is net rate of 'Red' diffusing to electrode = net rate of 'Ox' diffusing away:

$$t > 0\ x = 0\ \ln([\mathrm{Red}]_{x=0}/[\mathrm{Ox}]_{x=0}) = (-nF/RT)(E - E_{1/2}^{0})$$

that is the Nernst equation for a reversible system (i.e. fast electron transfer) going from $E < E_{1/2}^{o}$ to $E > E_{1/2}^{o}$.

As the potential is swept towards positive potentials the concentration of the reduced form at the electrode ($[\mathrm{Red}]_{x=o}$) decreases creating a concentration gradient at the electrode and the current:

$$I = n\,F\,A\,D_{red}\,(d[\mathrm{Red}]/dx)_{x=0}$$

Once $[\mathrm{Red}]_{x=0} = 0$ then $d[\mathrm{Red}]/dx$ is at a maximum and the (Nernst) diffusion layer continues to thicken and $d[\mathrm{Red}]/dx$ decreases.

Mathematical treatment of this equation is beyond the scope of this book but the main features of the voltammogram are (depending on sweep direction) the peak potentials ($E_{p,a}$ and $E_{p,c}$) and currents ($I_{p,a}$ and $I_{p,c}$), and $E_{1/2}$, the so-called half-wave potential.

5.2.8 **Cyclic voltammetry where both oxidized and reduced forms of a redox active species are soluble**

Cyclic voltammetry is a central technique in electrochemistry and is an extension of linear sweep voltammetry. The experiment is essentially identical in fact; with a linear (sweep) voltage ramp (i.e. at a particular scan rate) from the initial potential to a maximum potential (V_1, the switching potential) at which point the direction of the voltage ramp is reversed and the ramp continued until a second (switching) potential (V_2) is reached. The scan direction is reversed again to complete one cycle. A cyclic voltammogram recorded over a potential range which includes the half-wave ($E_{1/2}$) potential of a redox active species, is shown in Fig. 5.1.

Although the forward wave can be understood as a linear sweep voltammogram, the return cyclic from V_1 to V_2 involves a current in the opposite direction. Above $E_{1/2}^{o}$ the current will be determined by the slope of the concentration gradient (d[Red]/dx) that is the concentration in the bulk divided by the depth of the stagnant layer (fully developed Nernst diffusion layer). However, at all times the concentration of the oxidized form ([Ox]) in the diffusion layer is increasing towards the limit [Ox] = $[Red]_o$ and as the overpotential decreases (i.e. during the second sweep towards negative potentials) and approaches $\eta < +100$ mV, then the ratio of [Red]/[Ox] at the electrode will change (increase) and the gradient (d[Red]/dx) will decrease causing the current to fall. When the overpotential becomes negative (e.g. $\eta > -100$ mV) then the current will flow in the opposite direction as it will depend on the [Ox] in the stagnant layer in a similar way that current depended on [Red] at positive overpotentials and will reach a peak current when the gradient $(d[Ox]/dx)_{x=0}$ is at a maximum. Thereafter, the current will fall back to zero as the concentration of the oxidized form decreases since there is no oxidized form in the bulk to replace it.

The peak currents for reversible systems $I_{p,a}$ and $I_{p,c}$ are given by the **Randles-Sevĉik expression:**

$$i_p = 0.4463\,(F^3/RT)^{½}\, n^{3/2}\, A\, D_o^{½}\, v^{½}\, C^*$$

which at 2°C simplifies to

$$I_{p,a} = -I_{p,c} = (2.69 \times 10^5)\, n^{3/2}\, A\, D_o^{½}\, v^{½}\, C^*$$

If the electrochemical system under examination is reversible then the peak currents:

i) should be equal in magnitude: $i_{p,a} = -i_{p,c}$,

ii) should be increase with the square root of the scan rate, and

iii) potentials ($E_{p,a}$ and $E_{p,c}$) should be independent of the scan rate.

5.2.9 **Electrochemical and chemical reversibility**

The reversibility of an electrode reaction has a substantial impact on the appearance of the voltammograms obtained. Up to now, we have only considered redox couples that are reversible, however, many systems show irreversible electrochemical behaviour. Irreversibility will result in a change to the peak current, and, depending on the (redox)chemistry of the product, the appearance of additional redox waves in a cyclic voltammogram. There are different causes of irreversibility, however, each of which have their own impact on the general shape of a voltammogram.

Although we think of reversibility in a chemical sense (i.e. the molecular structure is retained or not during the redox process, see Box 5.3), the term electrochemical irreversibility does not imply a chemical reaction is taking place. Indeed, a redox reaction is electrochemically irreversible if the rate of heterogeneous electron transfer is too low for a Nernst equilibrium to be established at the electrode at all times (an assumption we made in the previous sections). For example, a significant deviation of α from 0.5 can cause irreversibility (see Chapter 4). However, even perfectly symmetric systems ($\alpha = 0.5$) can show irreversible behaviour if k^o is low; for example, where a redox active site is buried deep within

Box 5.3 Electrochemically driven chemical reactions

In earlier sections, the general shape of cyclic voltammograms of compounds that are chemically stable in both oxidation states (e.g. $[Fe(III)(CN)_6]^{3-}/[Fe(III)(CN)_6]^{3-}$) was considered. Cyclic voltammetry of compounds that undergo chemical reactions after oxidation or reduction can provide extensive mechanistic insight, especially if the rates of these reactions are similar to the timescale of the voltammogram (0.1–10 s). There is a broad range of classes of chemical reactions that can follow a change in oxidation state and a shorthand approach is used to describe these reaction sequences. Although many electrochemical reactions have analogues in reactions initiated by chemical oxidation, an important difference is spatial localization; that is, the reactions take place within the Nernst diffusion layer. A simple example is shown in diagram in Fig. 5.13 in which the species undergoing oxidation at the electrode (at the IHP/OHP) is not the same species as is present predominantly in the bulk solution.

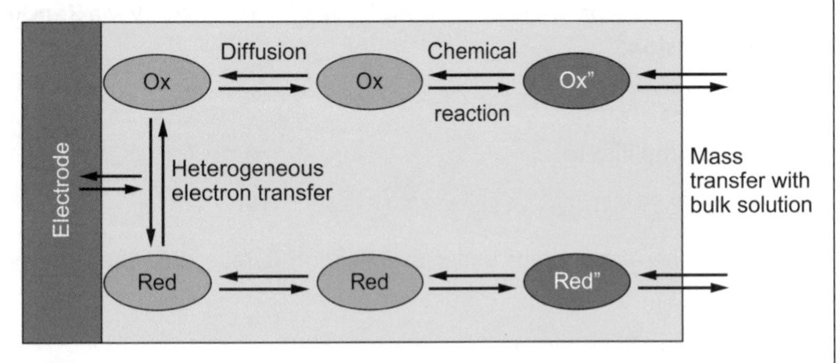

Figure 5.13 An important chemical component of an electrochemical reaction is where the redox active species present in the bulk solution is in equilibrium with another species that actually undergoes electron transfer at the electrode.

Each electrochemical (redox) step is labelled 'E' and every chemical step is labelled 'C' in a reaction scheme. For example, in an EC mechanism an electron transfer (E), for example, oxidation of $A \rightarrow A^+$, is followed by a chemical reaction (C) to form a new species (e.g. $A^+ \rightarrow B + H^+$). The properties of the product determine whether or not additional redox processes are observed and the overall shape of the voltammogram.

The oxidation of amines and heteroaromatics is an important field of synthetic organic electrochemistry as one-electron oxidation can be viewed, formally, as removal of an electron from the nitrogen's lone pair. Such oxidation often triggers further chemical reactions that can be of synthetic value. The organic amine dimethylaniline (DMA) is an example where other resonance structures of the monocation should be considered also. Electrochemical oxidation of DMA generates a radical cation in which the 'hole' is delocalized over the aryl ring as well as the amine nitrogen, represented as a series of resonance structures. The resonance structures show that the carbon at the para position shows radical character. The cyclic voltammetry of DMA between the OCP and positive potentials shows a chemically irreversible oxidation ($I_{p,c} = 0$) at around 1.1 V versus SCE and, on the return sweep to negative potentials, two reversible redox waves due to a product formed after oxidation of DMA (Fig. 5.14).

If we consider the mechanism for the reactions occurring, the first step is electrochemical (E). The concentration of DMA^+ at the electrode is the same as the concentration of the DMA in the bulk. The radical character at the para position is sufficient for formation of a C–C bond between the two aryl rings (C), which is stabilized

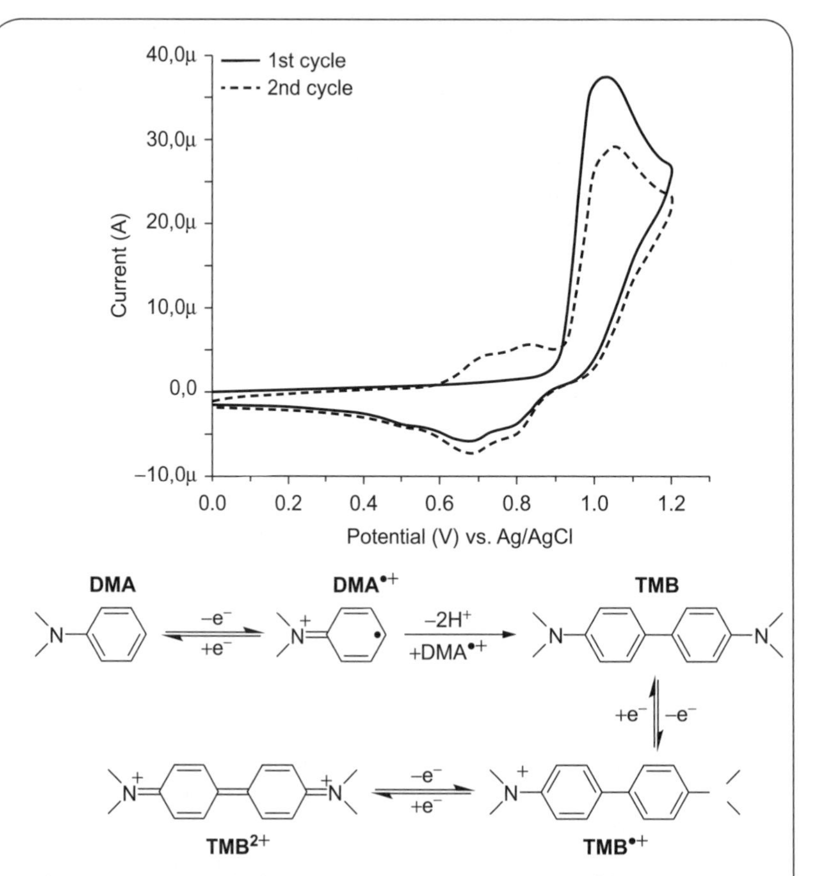

Figure 5.14 An example of an irreversible cyclic voltammogram of, for example, DMA (see the text). In this case oxidation at 1.0 V leads to formation of a radical cation (DMA^{+}) that undergoes dimerization by aryl-aryl C-C bond formation and deprotonation to form a new redox active species that undergoes oxidation at less positive potentials. This species remains with in the diffusion layer and on the second cycle undergoes oxidation and reduction also. This is an example of a chemically irreversible electrochemical reaction that is electrochemically reversible. Each electrochemical step is denoted 'E' and each chemical step 'C'.

by spontaneous loss of two protons (C) yielding the stable neutral redox active compound *N,N,N',N'*-tetramethyl-benzidine (DMB). However, at the potential required to oxidize the monomer, the DMB formed undergoes immediate two electron oxidation also to its dicationic form (DMB^{2+}) (E), and upon cycling towards less positive potentials, the two one-electron reduction waves of DMB^{2+} (EE) are observed.

Overall, the cycle voltammetry of DMA is described by a ECCEEE mechanism! If, however, we add a methyl group at the para position of DMA, the coupling step is prevented and a simple reversible redox wave is observed. The mechanism for DMA dimerization, although relatively complex, is the seen in many redox active aryl based systems and, especially, in oligothiophene and oligopyrrole electrochemistry.

In cyclic voltammetry the boundary between reversible and irreversible is somewhat blurred and the term quasi-reversible is used if ΔE is between 60 and 120 mV, provided $I_{p,a}/I_{p,c} = 1$ and $\alpha \approx 0.5$. >

Reversible $k^o >> 0.3\ v^{1/2}$ cm s^{-1}, quasi-reversible $0.3\ v^{1/2}$ cm s^{-1} $> k^o > 0.3\ v^{1/2}$ cm s^{-1}, irreversible $k^o << 0.3\ v^{1/2}$ cm s^{-1}.

a protein and the minimum distance from the electrode is greater than the thickness of the Helmholtz planes. In the these cases, the voltammetry will be reversible at low sweep (scan) rates and become irreversible ($E_{p,a}$ and $E_{p,c}$ will change, etc.) as the scan rate is increased.

The term quasi-reversible electrochemical system is used frequently and refers to a situation where the behaviour is dependent on the rate constant for heterogeneous electron transfer (k^o). At low scan rates the cyclic voltammogram will appear normal. As the scan rate is increased, the system will not be able to respond fully to the change in potential (i.e. electron transfer, and not diffusion, is rate limiting) and characteristic distortions.

In the case where $\alpha \neq 0.5$, we should remind ourselves of the Butler–Volmer equation, which contains the term α.

$$k_{red}^{et} = k^o exp\left(\frac{(1-\alpha)\left[E - E_F^o\right]nF}{RT}\right)$$

If the deviation is sufficiently large then the rate of back electron transfer is slow and we can will develop an alternative form of the Randles–Sevĉik equation, which includes the term α. The potential at which the peak currents are observed are shifted to more positive ($I_{p,a}$) or less positive ($I_{p,c}$) potentials.

$$I_{p,a} = \left(2.99\times10^5\right)n\left(\alpha n_a\right)^{1/2} D_A^{1/2} v^{1/2}\left[A\right]_o$$

where n_a is the number of electrons before the rate limiting step.

Chemically irreversible redox reactions can be much more complex. They are observed when product of the oxidation or reduction undergoes a chemical reaction to form a new species, or where there is an equilibrium between two species prior to oxidation/reduction. However, the rates must be similar or greater than the scan rate used in a voltammetric experiment in order to see an effect on the voltammetry.

5.2.10 **Practical aspects in cyclic voltammetry**

Typically, a cyclic voltammogram is initiated (initial potential) at the OCP since at this potential the ratio of oxidized and reduced species at the electrode is identical to their ratio in the bulk solution. The potential is held at the OCP for several seconds prior to the start of the voltage ramp to ensure an equilibrium situation is established. The scan rate can be from < 0.001 mV s^{-1} (limited by the effect of convection) to over 1000 V s^{-1} (limited only by the cells RC time constant (charging of the double layer) and the *iR* drop). The OCP is quite useful in indicating the ratio of oxidized and reduced forms of a redox active species in solution, for example, the cyclic voltammograms of $[Fe(II)(CN)_6]^{4-}$ and $[Fe(III)(CN)_6]^{3-}$ are identical except for the potential range at which the current is close to zero. For $[Fe(II)(CN)_6]^{4-}$ very little current flows at negative overpotentials as the oxidized form $[Fe(III)(CN)_6]^{3-}$ is not present in the bulk solution and vice versa. Although useful, more rigorous methods should be used such as those based on micro-electrodes (spherical diffusion conditions) and a rotating disc voltammetry approach.

5.3 General remarks

Although we have restricted our discussion here to simple step experiments, linear sweep, and cyclic voltammetry, there are many step (such as differential pulse voltammetry) and sweep techniques (such as AC voltammetry) available (see the Further Reading section at the end of the book). When the redox active species under consideration is chemically and electrochemically reversible then the analysis of the voltammogram is straightforward and characteristic potentials are easily related to each other. A few comments are worth making regarding these characteristic potentials, however, since a primary goal of voltammetry is to determine thermodynamic properties.

We should be aware of possible deviations between the half-wave potential, $E_{½}$, and the formal potential E_f^o including

- differences in the diffusion coefficient of the oxidized and reduced forms
- kinetics of electron transfer (activation barriers)
- solution equilibria
- irreversibility (in terms of electron transfer and in terms of chemistry)

In attempting to extract thermodynamic and kinetic data, especially rate constants, we should be mindful that the values obtained from fitting are only as good as the model that was used to generate the fit. It is often the case, especially where factors such as uncompensated resistance, differences in diffusion coefficients, change in pH at the electrode, and so on, influence the voltammetry and curve shape, that the model used provides an excellent fit but a meaningless result. In short, voltammetric techniques are at their most powerful as a tool to understand electrochemical systems when combined with other techniques and fitting is useful primarily when we already have a good understanding of the mechanisms involved,

5.4 Summary

This chapter should provide you with an understanding of:

- the various sources of current observed including Faradaic and non-Faradaic current
- the use of a three-electrode arrangement and the impact of cell resistance on the current measured
- the current response of an electrode following a potential step and during a potential sweep
- the concept of the Nernst diffusion layer and its development following a potential step
- the shape of voltammograms obtained with species present on the electrode and in solution
- the general features of a linear sweep or cyclic voltammogram
- the concept of electrochemical and chemical reversibility
- the occurrence of chemical as well as electrochemical reactions at the electrode

5.5 Exercises

5.1. How long will it take for a small molecule ($D = 5 * 10^{-5}$ cm^2 s^{-1}) and a large protein ($D = 1 * 10^{-7}$ cm^2 s^{-1}) to diffuse 100 μm away from the electrode? By what factor will the difference in diffusion coefficient be expected to affect the peak current ($I_{p,a}$), assuming that scan rate and concentration are the same in each case?

5.2. Justify the use of the equation $I_p = \frac{n^2F^2}{4RT} v A \Gamma_o^*$ to calculate peak current observed in a cyclic volltamogram of a species immobilized on the surface of an electrode (see text for details) by differentiating the following equation by parts:

$$I = nFA\frac{d\Gamma_{ox}}{dt} = \frac{n^2F^2 v A \Gamma_{ox}}{RT} \frac{\exp\left(\frac{nF(E - E^{0'})}{RT}\right)}{\left(1 + \exp\left(\frac{nF(E - E^{0'})}{RT}\right)\right)^2}$$

5.3. The following data set was obtained from cyclic voltammetry at a planar glassy carbon electrode (3 mm diameter) for a compound in solution, which undergoes fully electrochemically reversible one-electron redox chemistry. The diffusion coefficient for both oxidized and reduced forms is the same ($3.2 * 10^{-5}$ cm^2 s^{-1}) From the data, recorded at 265 K in the table:

(a) calculate the unit area capacitance of the electrode and

(b) determine the concentration of the reduced form of the compound.

Scan rate (V s^{-1})	$I_{p,a}$	
	at 0.10 V	at 0.56 V
	in μA	in μA
0.02	0.7	1.5
0.05	1.8	4.0
0.075	2.7	10
0.125	4.4	20
0.2	6.9	30
0.5	13.8	65
0.75	18.2	84
1.0	23.7	106
2.0	43.7	176
5.0	109.2	335
10.0	200.2	535

5.4. A famous organic chemist had trouble with a reaction to make an important intermediate in a 32-step total synthesis. He tried every chemical method he could think of and out of desperation asked an electrochemist he knew for help. The electrochemist after recording a cyclic voltammogram of the compound carried out a bulk electrolysis (i.e. oxidized everything in solution using a large area electrode) and then purified the compound by extraction and washing. The ^{1}H NMR spectrum looked largely similar to the starting material shown in the figure, except the coupling pattern in the aromatic region was simpler and he was missing a hydrogen.

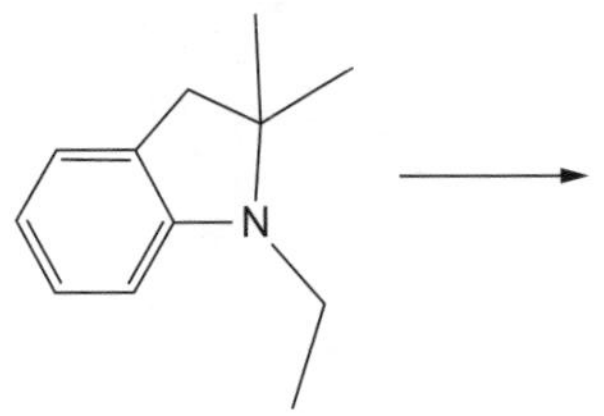

(a) Discuss the mechanism for the reaction that had occurred showing the individual electrochemical and chemical steps that are involved.

(b) Sketch the cyclic voltammogram you would expect the electrochemist recorded and indicate which processes occur at which potentials.

6 Batteries, fuel cells, and electrochemical impedance spectroscopy

6.1 Introduction

Batteries and fuel cells are sources of direct current and are, to all intents and purposes, the Galvanic cells we met in Chapter 2. Batteries differ from fuel cells in two important ways;

i) batteries have a finite amount of material that can be oxidized and reduced whereas in fuel cells fresh reactant is delivered on a continuous basis.
ii) catalysts are employed in fuel cells to increase electron transfer rates to useable levels (i.e. methanol or glucose oxidation), whereas in batteries this is not normally necessary since the reactants have low barriers to electron transfer (i.e. reactive metals such as lithium).

Probably the best-known batteries are the ubiquitous lead storage battery used in, for example, cars, the lithium ion battery in mobile phones and laptops, and the dry cell batteries used in practically every small portable electronic device from calculators and bicycle lights. Although many battery designs are in use today, they share characteristics that are essential in any, at least consumer, application. In this chapter, we will give a brief overview of battery and fuel cell designs and thereafter explore the technique of electrochemical **impedance** spectroscopy (EIS), which is a central tool in studying the properties and behaviour of such devices.

6.2 General issues in battery and fuel cell design

The basic components in a battery are the anode, cathode, electrolyte, a reductant, and an oxidant. The components determine the cell EMF and the internal resistance both of which limit the power provided by the battery (see Chapter 1). The cell EMF should be as high as possible to provide maximum power ($P = IV$); however, if it is too high then side reactions such as electrolyte oxidation can occur. More importantly ion mobility is important to prevent concentration polarization (internal resistance) from building up. The duration with which a battery can provide a useable voltage is limited primarily by the number of moles of anode and cathode reactants and the maximum current by the surface area of the electrodes and mobility of the redox active species.

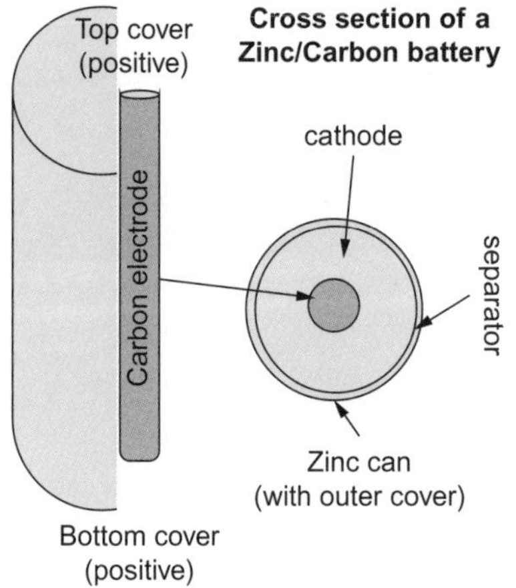

6.2.1 **Characteristics of an ideal battery and factors considered in battery design**

i) **High cell EMF**—the higher the EMF, the more work that can be done with the same number of electrons and hence moles of reducible and oxidizable material. The anode and cathode reactions are matched to achieve the maximum EMF that a particular electrolyte will tolerate and is typically 1.2 −1.6 V. A higher EMF is achieved by connecting several cells in series (three 1.2 V batteries in series give an EMF of 3.6 V).

ii) **Duration** (mA h) with respect to mass (kg)—a high power density is ideal such that each mole of electrons is carried by a minimum mass of material (lithium is therefore better than Cerium!). Although the ideal oxidizable material is H_2 in the metallic state, this is somewhat impractical and hence lithium (7 amu per electron) and aluminium (26/3 or 8.67 amu per electron) are highly suitable since they provide similar charge densities.

iii) **Non-explosive/flammable**—does not burn or explode when in use, that is, generate heat when connected to a load due to cell (internal) resistance (e.g. due to mass transport limitations/polarization), short circuits inside the battery or the formation of gases (e.g. H_2 in a lead-acid battery).

iv) **Stable electrolyte**—the electrolyte is one of the most challenging components as it must not be corrosive (i.e. react with the electrodes or with the battery/fuel cell casing). Fuel cells present an additional problem when operated at high temperatures (up to 900 K!). If a solvent decomposes it can release gases, such as CO_2 or H_2 that can lead to explosions.

v) **Non-toxic and sustainable**—does not contain components that are scarce or when released are toxic to humans or harmful to the environment (e.g. cadmium, mercury etc.).

vi) **Stable separator**—It is essential that the reductant and the oxidant do not react together directly as this would generate heat and not electrical current.

6.2.2 **Polarization**

In batteries, high ion mobility is essential to prevent concentration polarization (the development of a concentration gradient at the electrode). Polarization occurs when ions do not move quickly enough when current is drawn from the battery; that is, a net positive charge builds up at the solution near the anode and a net negative charge builds up near the cathode. The electrode areas are important also as the rate at which electrons can be transferred to and from the electrodes limits the maximum current that can be drawn. The rate constants (see Chapter 4) will be different for each electrode but this problem can be overcome by tuning the surface area of each electrode so that the overall rate at which electrons are taken up by the anode match the rate they are given up by the cathode.

6.2.2.1 Why do you always change both batteries even when only one is drained?

If two batteries are in series and one is nearly at the end of its useful life (i.e. nearly discharged or drained) then the cell voltage and hence current that it can be provide will decrease. A problem arises if it is in series with a new cell that can pass a much higher current since the maximum current will be limited by the drained battery and, worse still, the good battery will force some additional current through the drained battery causing heating (and possibly a rapid detrimental increase in temperature and pressure in the battery).

6.3 Common battery designs

Although numerous battery designs are in use, a brief overview of common cell designs is useful in order to draw comparisons with simple Galvanic cells and highlight the interdisciplinary nature of the teams needed to design batteries.

Not all batteries are similar to Galvanic cells. Organic polymer batteries use conducting polymers (e.g. polyacetylene, etc.) and, although polarization is normally avoided, organic polymer batteries store energy using electrode polarization. Two thin films of polyacetylene form the anode and the cathode, with charge making the anode become negative, taking up cations, and the cathode becoming positive and taking up anions. The direct current provided during discharge is due to the driving force (ΔG) for both polymer modified electrodes to depolarize.

6.3.1 Dry cell batteries

Dry cell batteries, used frequently in watches and calculators and so on, use a paste electrolyte, which is either acidic or basic. The first acid dry cell battery, by le Clanche (1839–1882), delivered an EMF of 1.5 V and was composed of Zn metal and a graphite rod (anode and cathode, respectively) separated by a paste containing MnO_2, NH_4Cl, and carbon.

The anode reaction is: $Zn^{2+} + 2e^- \rightarrow Zn(s)$

and the cathode reaction is: $2NH_4^+ + 2MnO_2 + e^- \rightarrow Mn_2O_3 + 2NH_3 + H_2O$

In alkaline variants of the dry cell battery, the NH_4Cl is replaced with KOH or NaOH

The anode reaction is: $ZnO + H_2O + 2e^- \rightarrow Zn(s) + 2OH$

and the cathode: $2MnO_2 + H_2O + 2e^- \rightarrow Mn_2O_3 + 2OH$

The alkaline variant is preferred as it has a longer shelf life due to the slower rate of Zn corrosion. Two other common batteries are the silver and the mercury cells, in which Ag_2O and HgO, respectively, replace MnO_2.

6.3.2 Rechargeable batteries

Recharging discharged batteries necessitates the reversal of the electrode reactions; that is, running the cell in an electrolytic rather than Galvanic mode. The materials and electrode reactions employed in the battery are not always easily reversed for any number of reasons. Typically, however, batteries that use pairs of electrodes of the second kind are suitable for recharging (e.g. the lead storage battery) if the products of each of the reactions adhere to the electrode then the reactions can be reversed and hence the battery recharged.

A commonly encountered rechargeable battery is the nickel-cadmium (NiCd) battery, which has as anode reaction $Cd(OH)_2 + 2e^- \rightarrow Cd(s) + 2OH^-$ ($E^0_{½}$ – 0.824 V) and cathode reaction: $NiO_2 + H_2O + 2e^- \rightarrow Ni(OH)_2 + 2OH^-$ ($E^0_{½}$ + 0.61 V).

6.3.3 **Lead storage battery**

The lead storage battery, developed originally by Planté in the 1860s, has achieved such importance mainly because it is useful, robust, has a long operational lifetime, good energy storage capacity, and the ability to 'recharge' the battery numerous times over several years is essential for its use in cars. Above all, however, this battery design (large electrode surface areas) can deliver high currents with low RC time constants, which is essential when starting a vehicle; think back to Chapter 1 and our discussion of the effect of current on the EMF of a battery. The lead storage cell delivers approximately 2 V and typically six cells are connected in series to produce 12 V of direct current (DC). During discharge (i.e. use), the H_2SO_4 is consumed to form solid $PbSO_4$, which coats both the anode and cathode. This results in a decrease in the density of the electrolyte to less than the optimal 1.28 g/cm^3; the density of the electrolyte can therefore be used to monitor the extent of discharge (how much current the battery can still deliver).

A plate of lead serves as the anode and a lead plate coated with a layer of lead oxide (PbO_2) the cathode. The electrolyte is concentrated sulfuric acid (H_2SO_4).

The anode reaction is Pb(0) to Pb(II):

$$PbSO_4(s) + H^+ + 2e^- \rightarrow Pb(s) + HSO_4^-(aq) \qquad E^o = -1.685\ V$$

and the cathode reaction is Pb(IV) to Pb(II):

$$PbO_2(s) + HSO_4^-(aq) + H^+ + 2e^- \rightarrow PbSO_4(s) + 2H_2O \qquad E^o = -0.356\ V$$

The overall cell reaction is:

$$Pb(s) + 2HSO_4^-(aq) + PbO_2(s) \rightarrow 2PbSO_4(s) + 2H_2O \qquad E_{cell} = 2.041\ V$$

The lead and lead salts involved are all insoluble and are mechanically stable so that the electrodes do not change in overall lead content regardless of the extent of discharge of the battery. Two further advantages are that all components are solids (except of course for the sulfuric acid) and hence the cell voltage is essentially constant even when nearly fully discharged, and a single cell chamber can be used (i.e. no membranes, etc.). The lead storage battery is recharged by forcing current in the opposite direction thereby reversing both reactions. However, the reverse reaction can result in electrolysis of water to produce H_2 and O_2. This means that the battery has to be 'topped off' occasionally with water (i.e. water added).

This feature of lead storage batteries has a consequence that when a vehicle with a 'dead' battery is given a 'jump-start', in ignition of a mixture of H_2 and O_2 gas that can form in the battery leads to an explosion.

6.3.4 **The lithium ion battery**

The lithium ion battery, developed in the early 1980s, is perhaps the best-known battery both because it is light and has a high power density, but also because it has the unfortunate potential to get hot and burst into flames at inconvenient moments. The anode and cathode materials are $LiCoO_2$ and carbon with a liquid electrolyte (typically ethylene carbonate) and perforated plastic films to separate the electrodes that are wrapped in a spiral arrangement. The key aspect in the performance of these batteries is that lithium can move rapidly from one electrode to the other and can intercalate (move in between layers of carbon or CoO_2), which prevents the build-up of a passivating layer on the surface of the electrodes.

6.4 Fuel cells

Davy's fuel cell: carbon | nitric acid, H_2O| O_2 | carbon.

The reactions used to drive the fuel cells used on the NASA Space Shuttles are:

Anode: $2H_2(g) + 4OH^-(aq) \rightarrow 4H_2O(l) + 4e^-$

Cathode: $O_2(g) + 2H_2O + 4e^- \rightarrow 4OH^-(aq)$

Overall reaction $2H_2(g) + O_2(g) \rightarrow 2H_2O(l) + \text{energy}$

Although much less frequently encountered in daily life, fuel cells are not a recent invention. At their simplest, a fuel cell is a Galvanic cell in which the reactants are supplied continuously, that is, a chemical oxidant and reductant are replenished continuously. Humphrey Davy built the first fuel cell based on carbon electrodes, nitric acid (aqueous) and oxygen, which had a low, essentially useless, EMF. The first practical fuel cell was developed in 1839 by William Grove, which was essentially the same electrolytic cell used to split water into hydrogen and oxygen gas operated in reverse.

Although, the internal combustion engine seems to have stifled the development of fuel cells in the twentieth century, in reality, the challenges faced in developing fuel cells that can operate with cheap materials (electrodes/electrolyte) safely and with high efficiencies are substantial in terms of the materials science, catalysis, and engineering involved. Indeed, NASA's space programme in the 1960s provided the stimulus to make a huge step forward in the development of fuel cells.

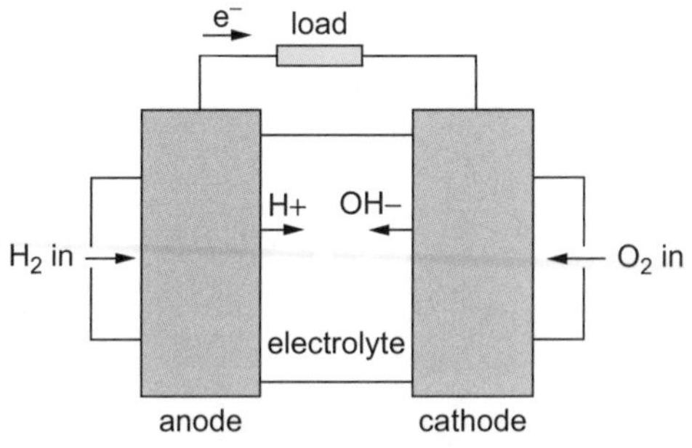

The reactions in a gas feed fuel cell occur at the interface between the electrode, the electrolyte, and the gas. Because of this, electrodes are designed to maximize the surface area and only thin films of electrolyte lie between the electrode and the sparingly soluble gas.

A major challenge in gas fuel cells is the boundary between the gas, liquid, and solid phases. This meeting point determines performance and how the best use can be made of electrochemical catalysts (electrocatalysts). Furthermore, the electrolyte used depends on the nature of the oxidant and reductant used. Cells that operate at room temperature typically have liquid electrolytes, but high temperature fuel cells often have solid electrolytes, such as certain phases of aluminium oxide or a polymer electrolyte.

Thermodynamics of fuel cells and batteries

In contrast to combustion (1200–1800 K), reactions in fuel cells are carried out at low temperature (300–900 K) and hence entropy contributions to Gibbs energy are significantly reduced. For example, methane can be burned to generate heat but the reaction can also be carried out in a fuel cell to deliver the energy as direct current. Furthermore, while direct burning of methane to produce heat is an important reaction, burning H_2 gas is not useful as it burns explosively (it knocks).

$CH_4(g) + 2O_2(g) \rightarrow CO_2(g) + 2H_2O(l) + \text{energy}$

The energy can be released in a more controlled manner as electricity using a fuel cell. Consideration of the thermodynamics of the reaction allows us to understand why energy can be released more effectively using a fuel cell by comparing it with an ideal heat engine. The electrical work done is related to the

Gibbs energy change and can be expressed in terms of the enthalpy and entropy change in the overall chemical reaction:

$$w_{elec} = -\Delta G = -(-\Delta H - T\Delta S) = -\Delta H\,(1 - T\Delta S/\Delta H)$$

The rearranged equation shows that if ΔS is negative for a reaction, then at low temperatures the enthalpy change is nearly fully available and as the temperature increases the amount of work that can be done decreases. In contrast for combustion (which is at a much higher temperature) the amount of work done can be calculated using the concept of a reversible heat engine operating between T_h and T_c, where ε is the maximum efficiency:

$$w_{thermal} = q_{total}\varepsilon = -\Delta H\,(T_h - T_c)/T_h$$

Burning releases heat (q_{total}) of which only a part is useable to do work.

For a lead (acid) storage battery:

$$\Delta G^o = -376.97\ \text{kJ mol}^{-1},\ \Delta H^o = -227.58\ \text{kJ mol}^{-1},\ \Delta S^o = -501.1\ \text{J mol}^{-1}\ \text{K}^{-1}$$

Assuming $T_h = 600$ K and $T_c = 300$ K and that the battery operates at 300 K, the calculation leads to the result that $w_{elec}/w_{thermal} = 3.1$. This shows that more work can be done by the electrochemical reaction than by the thermal reaction.

$W_{thermal}$ is the heat put into the system, T_c and T_h are the temperatures of the hot and cold reservoirs, respectively. **Carnot's theorem** states that a Carnot engine has the maximum efficiency possible when operating between to heat reservoirs.

6.4.1 **The direct methanol fuel cell (DMFC), catalysts for water oxidation, and CO poisoning**

Methanol is an attractive alternative to H_2 as a chemical fuel as it is bioavailable and has a relatively high energy content per unit mass. The major challenge in using it is that direct electron transfer from the electrode to methanol is not possible but instead specific adsorption and then reaction of the compounds with surface atoms on the electrode is required, involving a complex series of chemical reactions from CH_3OH to CO_2. Hence, the electrode material must be chosen to lower the activation energy, and thereby the overpotential needed (Chapter 4), for all steps. Developing an electrode catalyst resilient to poisoning by intermediates, for example, CO, which is an excellent ligand for many metals and can result in a reduction of catalytic activity, is therefore a key challenge. For the cathode reaction (reduction of oxygen to water) a further complication is the formation of hydroxyl radicals, superoxide and hydrogen peroxide, which can corrode fuel cell components.

Nature uses energy rich molecules such as glucose or polysaccharides to generate proton gradients and make electron equivalents (riboflavin and NADH), which can be oxidized at an electrode relatively easily. It is no surprise then that an important area of research in low temperature fuel cells is the use of bacteria and algae to metabolize complex carbohydrates anaerobically and to deliver the electrons to an anode in a so-called biofuel cell.

Anode:	$CH_3OH + H_2O \rightarrow CO_2 + 6H^+ + 6e^-$
Cathode:	$6\,H^+ + 3/2\,O_2 + 6e^- \rightarrow 3\,H_2O$
Overall:	$CH_3OH + H_2O + 3/2O_2 \rightarrow CO_2 + 3H_2O$

6.5 Electrochemical Impedance Spectroscopy (EIS)

A key challenge in the study of complex electrochemical devices is that batteries and fuel cells are largely inaccessible to spectroscopic techniques, especially when operated at high temperature. For example, the study of a high temperature gas/liquid fuel cell is complicated by both the lack of optical access and the fact that only a tiny fraction of the material from which they are constructed is actually active, for example, ternary sites in fuel cells where there is a common boundary between solid, liquid, and gaseous phases.

Potentiodynamic electrochemical methods, such as voltammetry, seem well suited to the study of these systems, and polarization (IV) curves are useful; for example, in fuel cells to determine cell resistance. The capacitances and impedances encountered in fuel cells, which typically have high electrode surface areas, poise a challenge. Techniques such as cyclic voltammetry cause large perturbations; that is, large currents flow, and polarization and iR drop interfere dramatically distorting the shape of the voltammogram obtained.

The inverse of impedance is **admittance**, that is, the ability of a circuit component to allow current to flow. Often, it is conceptually easier to think of differences in admittance rather than differences in impedance.

A key electrochemical technique used in studying batteries and fuel cells is EIS. EIS uses small amplitude oscillations in voltage to induce an alternating current (AC), but the small amplitude means that the system under study suffers minimal perturbation. In the EIS experiment a potential can be chosen at which no, or very little, Faradaic current is generated, for example, to study processes such as reorganization of the diffuse layer at an electrode. Of course, potentials at which Faradaic current flows (i.e. close to the $E_{½}$ of redox active species present) can be chosen also and in this case other processes become important such as mass transport. The main point is, however, that the small amplitude of the voltage oscillation means irreversible changes to the system can be avoided. The technique has seen extensive application in battery and fuel cell research and hence its discussion here is appropriate.

EIS is a remarkably simple technique experimentally. If we hold a system (e.g. an electrode) at, for example, 0.3 V and measure the current that flows, then it will be constant and equal to 0.3 V*R (the DC resistance). If we change the voltage by 10 mV then the double layer will reorganize in response and a current (non-Faradaic) will flow for a time due to movement of ions towards and away from the electrode and the changes in solvent dipole orientation. If we oscillate the potential by +/−10 mV, that is, vary the potential between 0.29 and 0.31 V sinusoidally, then the double layer will reorganize continuously but with a small lag in time in responding to the change in voltage. At low frequencies of oscillation, the current response will match the voltage change closely and plots of current and of voltage over time will overlay almost perfectly. As the frequency of oscillation, increases then the reorganization will not be able to occur as fast as the voltage is changing and the response will increasingly lag behind the change in voltage. At high frequencies, the voltage oscillation will be too fast for the solvent to respond at all and the current response will be negligible.

Polymer modified electrodes are a good example of a system in which the use of impedance spectroscopy is commonplace. As the potential varies close to the $E_{½}$ of the redox active units in the polymer film, then ions will migrate in and out of the film to compensate for changes in charge. As the frequency increases, the movement of ions is not fast enough to keep track with the change in redox state and hence the current response will not match exactly the change in applied voltage (a phase shift will become apparent).

By measuring the ability of the system to respond to a change in voltage, that is, 'generating an impedance spectrum', we determine two components to the impedance, which are often presented in complex plane notation in the form of a Nyquist plot in which a real (abscissa) and so-called imaginary (ordinate) component of the total impedance is presented.

6.5.1 **What is impedance (Z)?**

Before we discuss EIS, we should first consider what impedance (Z) is. The simplest definition of impedance is the retardation of the flow of charge (i.e. a reduction in current). A solid-state resistor is a well-known example of an impedance element. The current response of a resistor to both a direct and an alternating voltage is described by Ohm's law, $V = IR$ and it is notable that the current across a resistor is directly dependent on the potential and not the rate of change in potential. Hence, the current responds instantaneously to a change in voltage. The impedance created by a resistor is due to energy loss (dissipated as heat) as the current flows through it. The contribution of impedance due to energy loss is referred to as the *real* **component of the impedance** (Z_{Re}).

We will not consider the inductor in this discussion, however, its impedance is $Z = jL\omega$.

A capacitor will also introduce an impedance into a circuit in that it affects the flow of current, but, as we will discuss next, it does so in a more complex way. The impedance to current due to a capacitor has a fundamentally different origin than that of a resistor. If we apply a direct voltage across a capacitor then current will flow transiently until the capacitor is fully charged and afterwards the current will be zero. Hence, the impedance initially is zero and rises rapidly over a short time. The response of a capacitor to an alternating voltage depends on the frequency of the alternation and its contribution to impedance is due to energy storage and not energy loss. Its impedance is referred to as being purely imaginary, however, this is a mathematical concept ($j = \sqrt{-1}$) arising from phasor diagrams and not intended to imply that the impedance is not 'real' in the sense that it is not measurable—both real and imaginary contributions to the total impedance of a circuit are measurable.

In this section, we are concerned with the measurement of the total impedance of a circuit containing resistors and capacitors, which, as we will discover, is the sum of their real (energy loss) and imaginary (energy storage) contributions:

$$|Z| = \sqrt{Z_{Re}^2 + Z_{Im}^2}$$

The magnitude of the total impedance $|Z|$ depends on the frequency of the alternating voltage and the amplitude of the voltage oscillation. It is for this reason that the technique is **'spectroscopy'**.

Z_{Re} and Z_{Im} are often labelled Z' and Z''. In this chapter, we will use the former only for convenience.

6.5.2 **The EIS experiment**

AC impedance spectroscopy is a relatively simple experiment. The basis of impedance spectroscopy is the study of how the current responds to a change in potential (alternating voltage). We apply a harmonically oscillating voltage with an amplitude $\left(\frac{V_{max} - V_{min}}{2}\right)$ (typically +/– 5 mV), often superimposed on a direct voltage, and a frequency (f, or more usually angular frequency $\omega = 2\pi f$). The current response to this perturbation is recorded and the maximum current (I_{max}) and the extent to which the maximum of the current precedes or lags behind the maximum voltage (E_{max}) are determined. The maximum current (I_{max}) and the phase shift (ϕ) are sampled sequentially for each frequency over a wide range of frequencies (0.1 to 1000 000 Hz). Typically the data is represented as Bode plots of *log|Z(ω)| versus log(freq) and ϕ vs log(freq)*. The log and semi-log plots, respectively, allow for visualization of data over several orders of magnitude (Fig. 6.1).

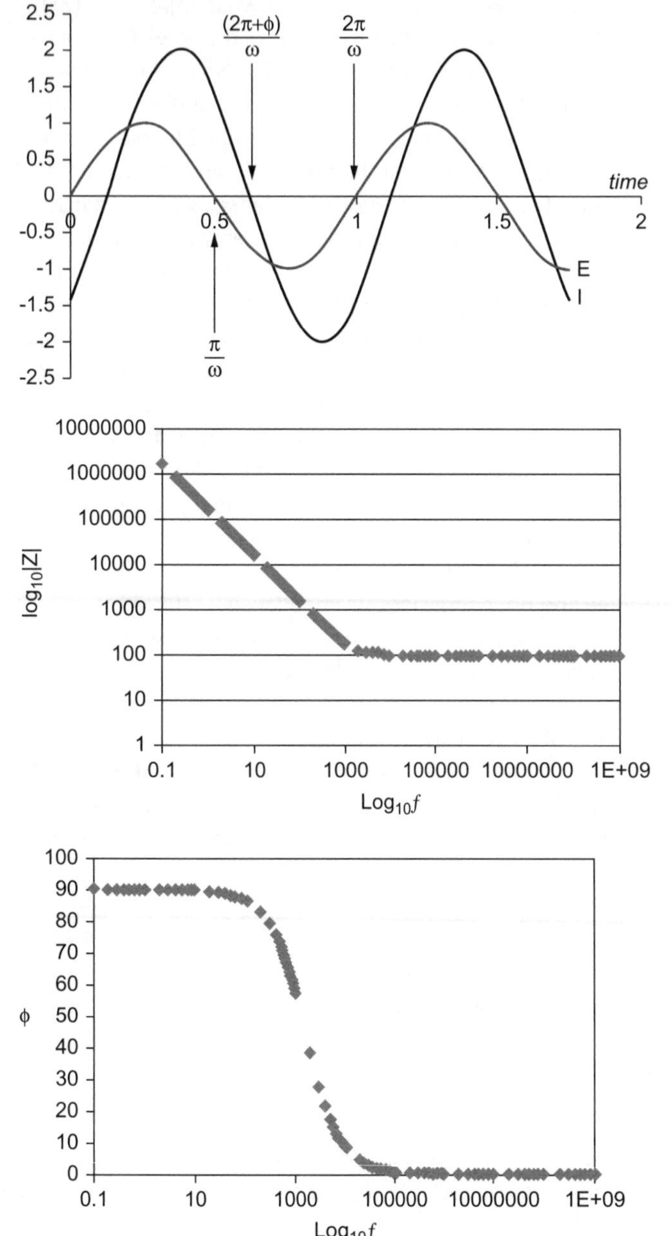

Figure 6.1 (Right) E vs time and I vs time plots showing I_{max} and f. (Left) Bode plots of log(Z) vs log(freq) and ϕ vs log(freq) for resistor and capacitor in series.

Box 6.1 Phasor representation of the time dependence of current and voltage

The potential and current at any time during a sinusoidal alternation are:

$$E_t = E_{max}\sin\omega t \text{ and } I_t = I_{max}\sin(\omega t + \phi)$$

where ϕ is the phase shift.
The impedance ($Z(\omega)$) at any time is given by Ohm's law and related to the total impedance $|Z|$ and the phase shift (ϕ):

$$Z(\omega) = \frac{E(t)}{I(t)} = \frac{E_{max}\sin(\omega t)}{I_{max}\sin(\omega t + \phi)} = |Z|\frac{\sin(\omega t)}{\sin(\omega t + \phi)}$$

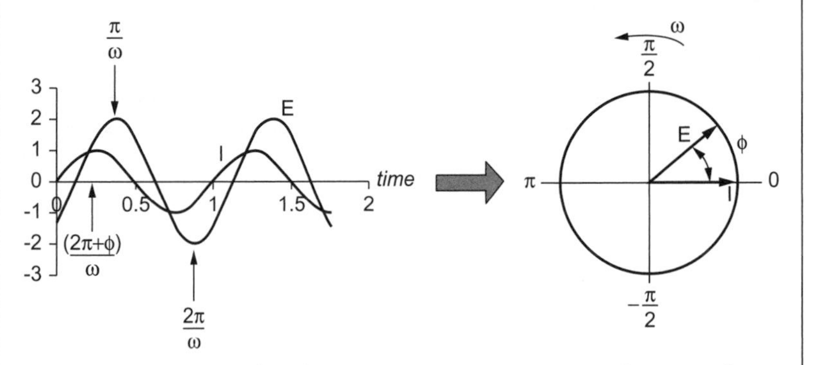

Figure 6.2 Phasor diagram (right) showing phase shift ϕ between the phasors E and I. In this case the current is taken as the reference point

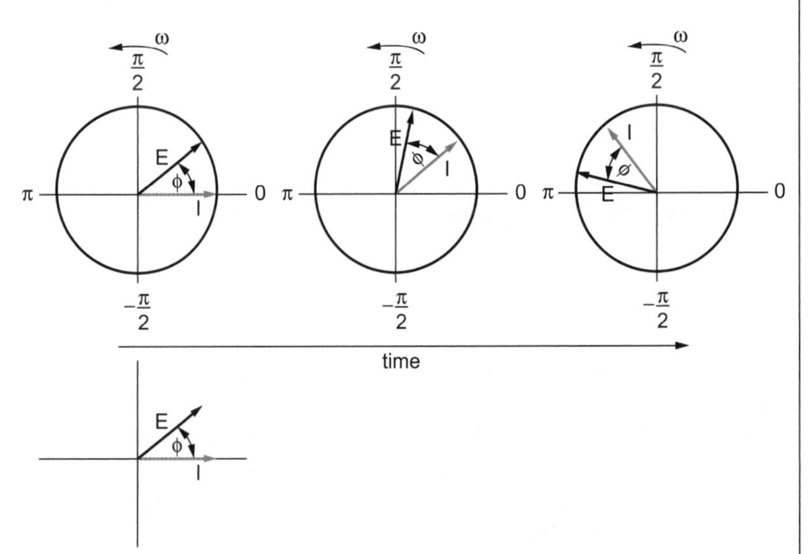

Figure 6.3 The phasor diagram showing how the voltage and current phasors evolve over time. Note that the phase shift is constant.

The time dependence of the current response to a sinusoidal alternation of potential can be represented as a phasor diagram (Fig. 6.3) instead of a plot to E versus time and I versus time ($\dot{E}$ and $\dot{I}$ represent voltage and current phasors, but we will use E_{max} and I_{max} for convenience). If we represent the E versus time and I versus time plots for a sinusoidal variation using a phasor diagram, by convention we take current as the reference. This representation shows that the total impedance $|Z|$ (i.e. E_{max}/I_{max}) is only experienced at all times when the phase shift (ϕ) is zero. When the phase shift is not equal to zero then we can use Euler's relation and complex plane notation (real and imaginary numbers) to determine the contribution of energy loss (real impedance Z' or Z_{Re}) and energy storage (imaginary impedance Z'' or Z_{Im}).

Current is taken as the reference for convenience, since for a capacitor the potential lags behind the current, see Fig. 6.2.

The point on the circle in the Argand diagram is always a distance Z from the origin, but the angle (ϕ) depends on what point in the cycle the system is at any particular time. The coordinates of that point are Z_{Re}, jZ_{im}, where $j = \sqrt{-1}$

The total impedance, Z, is calculated using basic trigonometric rules:

$$Z = Z_{Re} + jZ_{im} = |Z|(\cos\phi + j\sin\phi)$$

The imaginary number can be dealt with by making use of the Pythagoras theorem and taking the absolute square of the total impedance; in the complex plane notation this is:

$$Z^2 = (|Z|(\cos\phi + j\sin\phi))(|Z|(\cos\phi - j\sin\phi)) = |Z|^2(\cos^2\phi + \sin^2\phi)$$

Remember that the absolute square of the modulus in a complex plane is obtained by multiplying by the complex conjugate; that is, if $Z = a + jb$, then $|Z|^2 = (a + jb)(a - jb) = a^2 + b^2$.

and

$$Z^2 = (Z_{Re} + jZ_{im})(Z_{Re} - jZ_{im}) = Z_{Re}^2 + Z_{im}^2$$

Furthermore, the phase shift is related to the ratio of impedance due to energy storage and energy loss:

$$\phi = \tan^{-1}\frac{|Z_{Im}|}{|Z_{Re}|}$$

6.5.3 **The total impedance (|Z|)**

The total impedance ($|Z(\omega)|$) at each frequency is calculated using Ohm's law ($Z = E_{max}/I_{max}$) and is the sum of contributions from impedance due to energy loss (Z_{Re}) and impedance due to storage of energy (Z_{im}), where Z_{Re} is the 'real' and Z_{im} is the 'imaginary' impedance. The terms 'real 'and 'imaginary' are purely mathematical, however, and arise from the use of complex plane notation. Indeed, in analysing circuits, we want to extract contributions of energy storage and energy loss to the total impedance measured (Z) and the phase angle (ϕ).

Together, Z and ϕ provide polar coordinates for a complex plane called a Nyquist plot (shown in Fig. 6.4). In the Nyquist plot, the abscissa is the real (in phase) impedance (Z_{Re} or Z') and corresponds to the ability of the circuit to resist the flow of current (with energy lost as heat). The ordinate is the imaginary (out of phase) component (Z_{Im} or Z'') and is the ability of the circuit to store energy.

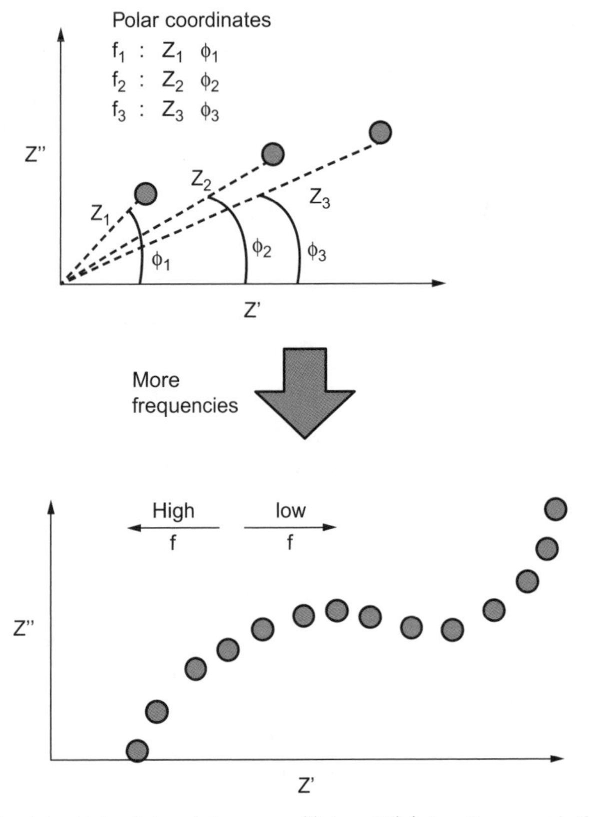

Figure 6.4 The data obtained at each frequency *(f)*, ϕ and $Z(w)=E_{max}/I_{max}$ provide the polar coordinates for each point on the plot. Note that in the plot obtained, the higher the frequency the closer the data point is to the origin.

The Nyquist plot can be handled mathematically as a complex plane (a + ib) with the abscissa as the 'real' impedance, and is related to the total impedance Z as:

$$Z_{Re} = |Z|(\cos\phi)$$

and the ordinate is the 'imaginary' impedance; for which we use the imaginary number $j = \sqrt{-1}$.

$$Z_{im} = |Z|(j \sin\phi)$$

The total impedance over a circuit can be calculated using Kirchhoff's laws:

$Z^* = Z_1 + Z_2 + Z_3 \ldots$ when in series and $\frac{1}{Z^*} = \frac{1}{Z_1} + \frac{1}{Z_2} + \frac{1}{Z_3} \ldots$ when in parallel (Fig. 6.5).

In reading a Nyquist plot (such as those shown in Fig. 6.4), the lowest intercept on the abscissa, which corresponds to the impedance at highest frequency, as we will see later, is the series resistance, while higher intercepts on the abscissa are additional resistances that are in parallel with capacitances.

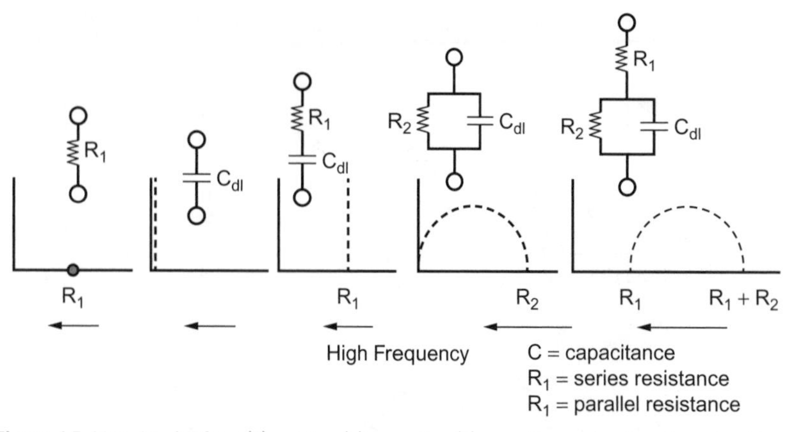

Figure 6.5 Nyquist plot for a (1) resistor, (2) capacitor, (3) capacitor and resistor in series, (4) resistor in parallel with a capacitor and (5) a resistor in series with a 'capacitor in parallel with a resistor'.

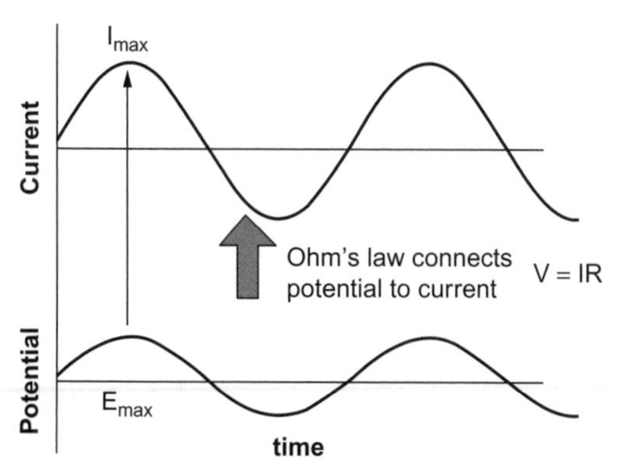

Figure 6.6 Potential and current response for a resistor.

6.5.4 **Impedance spectrum of a resistor**

A resistor responds instantaneously to a change in voltage (Figure 6.6) regardless of the frequency of the alternating voltage. Ohm's law can be used to calculate the current at any time ($V/R = I$) and the phase shift (ϕ) is zero. Hence, the impedance (Z) is the same at all times regardless of the point in the cycle. The impedance of the passage of current through a resistor does not result in the storage of energy ($Z_{Im} = 0$) but instead all energy is dissipated as heat (Z_{Re}).

Since $Z(\omega) = Z_{Re} + Z_{Im}$, then for a resistor, at all frequencies $Z(\omega) = Z_{Re} = R$ (Fig. 6.7) and since $\tan\phi = \frac{|Z_{Im}|}{|Z_{Re}|} = \frac{0}{R} = 0$ hence $\phi = 0$.

6.5.5 **Impedance spectrum of a capacitor**

The total charge (Q) that can be stored by a capacitor is dependent on the voltage: $Q = CE$.

Hence, if we change the voltage then the maximum amount of charge stored changes and current will flow until the new level is reached. If we change the potential in a continuous manner then current will therefore also flow continuously. In our analysis, we begin with the relation between current (I) and the rate of change of potential (dE/dt):

$$I = \frac{dQ}{dt} \quad \text{and} \quad I = C\frac{dE}{dt}$$

The current at any time can be calculated from the slope of the tangent of the graph 'E versus t' multiplied by the capacitance (C). The slope (and hence the current) is zero when E is at a maximum (E_{max}) and is greatest in magnitude when E is at the mid-point between the E_{max} and E_{min}. Hence, at the start of the experiment

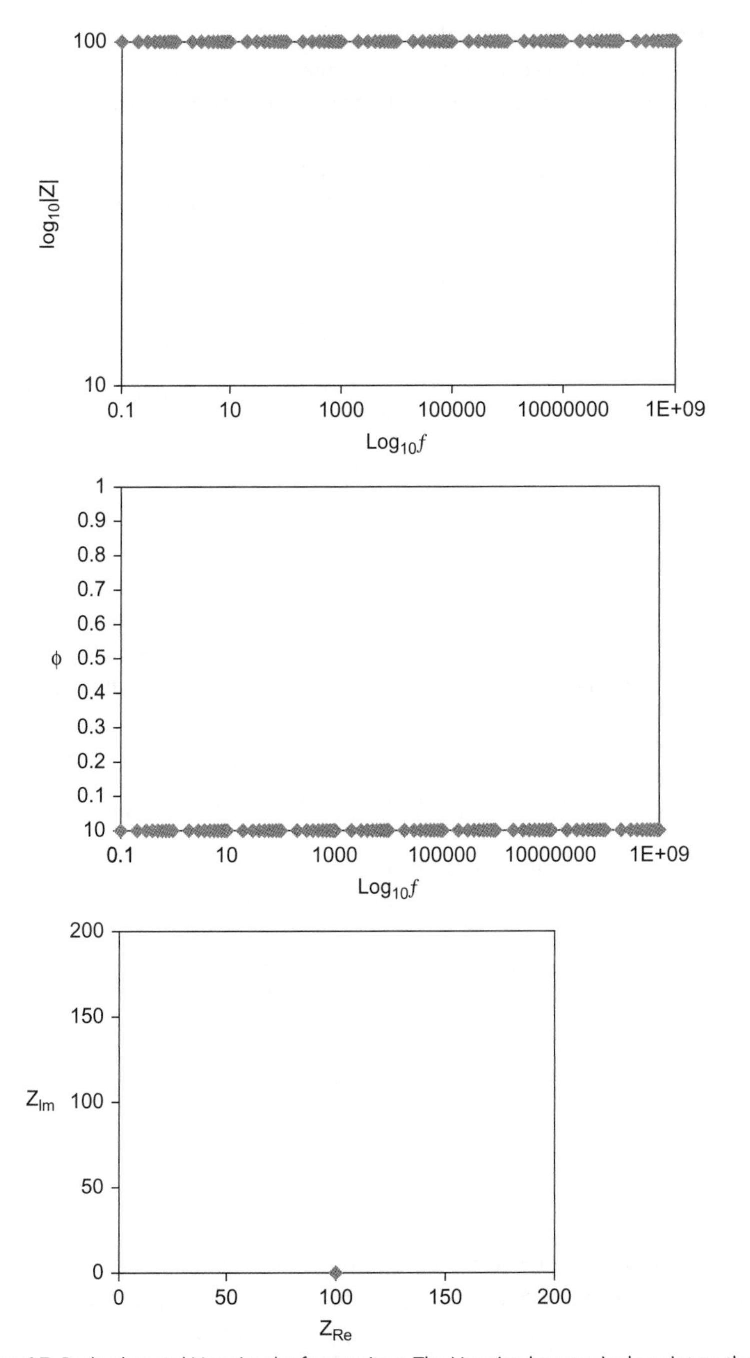

Figure 6.7 Bode plots and Nyquist plot for a resistor. The Nyquist shows a single point on the abscissa, in this case at 100 W, but in fact all data points fall on the same point.

the rate of change is greatest and hence the current is at a maximum. As a consequence, we say that the current 'leads' the voltage and the phase shift is π/2. Note that we take the convention that the phase shift to be positive as it is mathematically convenient.

We will now consider the time dependence of the current response to an alternating voltage. The voltage at any time is related to the frequency of alternation of the voltage:

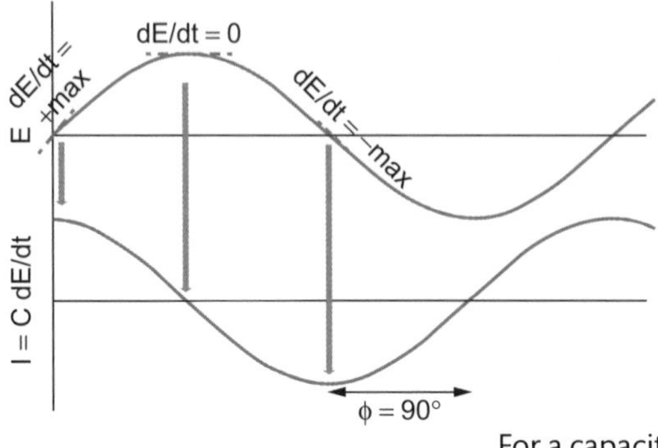

$$E(t) = E_{max} \sin \omega t$$

and hence the current at any time is given by:

$$I(t) = C\frac{dE(t)}{dt} = C\frac{dE_{max} \sin \omega t}{dt} = C\,E_{max}\frac{d\sin \omega t}{dt} = \omega C\,E_{max} \cos \omega t$$

Since $\cos \omega t = \sin\left(\omega t + \frac{\pi}{2}\right)$, then:

$$I(t) = \omega C\,E_{max} \sin\left(\omega t + \frac{\pi}{2}\right)$$

For a capacitor, the maximum current (I_{max}) flows when the potential is at the mid-point ($E_{max} - E_{min}$) (i.e. at $t = n\pi$), since the phase shift ϕ between E_{max} and I_{max} is 90° (π/2). The maximum current (I_{max}) is therefore at $t = 0$ and hence the impedance (Z):

$$|Z(\omega)| = \frac{E_{max}}{I_{max}} = \frac{E_{max}}{\omega C\,E_{max} \sin\left(\omega(0) + \frac{\pi}{2}\right)} = \frac{1}{\omega C \sin\left(\omega(0) + \frac{\pi}{2}\right)} = \frac{1}{\omega C \sin\left(\frac{\pi}{2}\right)}$$

However, the phase shift creates a problem as the voltage and current are, in this case, orthogonal phasors ($\dot{I}$ and $\dot{E}$, *which correspond to* I_{max} *and* E_{max}, *respectively*) and hence we have to express the phasors in complex notation ($a + jb$, where $j = \sqrt{-1}$). In other words, the sine of 90° in a Cartesian plan is 1, however, because the point lies on the imaginary axis (ordinate) of our complex plane, then it is $\sin\left(\frac{\pi}{2}\right) = \sqrt{-1} = j$ and hence:

$$|Z(\omega)| = \frac{1}{\omega C\, j} = \frac{-j}{\omega C}$$

and

$$\dot{E} = \frac{-j}{\omega C}\dot{I}$$

The frequency dependence of the response of a capacitor to a change in potential is quantified as the capacitive reactance (χ_c), which has units of ohms and its value decreases as frequency increases $1/\chi_c \propto f$. In other words, as the frequency is increased there is less and less time to fully charge the capacitor and hence to reach maximum impedance.

$$\chi_c = 1/\omega C \text{ and hence } \omega C = \frac{1}{\chi_c}$$

We can write the equation:

$$E_{max} = (-j\,\chi_c)\, I_{max}$$

where '$-j\,\chi_c$' is equivalent to the impedance of a resistor but it is also frequency dependent.

As the frequency decreases, its impedance ('imaginary resistance') increases. A capacitor affects the flow of current by storing energy and not dissipating it (i.e. $Z_{Re} = 0$) and hence:

$$|Z| = \sqrt{Z_{Re}^2 + Z_{Im}^2} = \sqrt{Z_{Im}^2}$$

and the phase angle is 90°:

$$\tan\phi = \frac{|Z_{Im}|}{|Z_{Re}|} = \frac{|Z_{Im}|}{0} = \infty$$

In summary, the impedance of a resistor to both alternating and direct current is the same and is due to the dissipation of electric energy as heat. In contrast, the impedance of a capacitor is due to energy storage and release through charging and discharging. The impedance to direct current is infinite except for a short period following a change in voltage whereas the impedance of AC is dependent on the frequency of the alternation (ω).

Impedance of a resistor: $Z(\omega) = R$

Impedance of a capacitor: $|Z(\omega)| = \sqrt{(-j/\omega C)^2}$

The difference in mechanism of impedance (energy loss (Z' or Z_{Re}) and energy storage (Z'' or Z_{Im})) requires the use of a complex plane ($a + jb$, where $j = \sqrt{-1}$) to determine the contribution of each impedance type to the total impedance (Z) of a circuit. In Sections 6.5.6 and 6.5.7 we will predict the total impedance of several circuits in terms of total impedance, using Kirchhoff's laws and the phase shift ϕ. In addition, we will consider how AC moves through a circuit at low, high, and intermediate frequencies.

At high frequencies a capacitor does not have time to charge fully (i.e. its capacitive reactance decreases) and as a result current flows as if it were a low impedance resistor.

6.5.6 **EIS of a series RC circuit**

The impedance of a resistor and capacitor connected in series (and RC circuit) can be calculated using Kirchhoff's law:

$$Z(\omega) = Z_1 + Z_2 = R - j/\omega C$$

We will first consider how low and high angular frequency (ω) can simplify the equation. At low frequency, the second component ($-j/\omega C$) becomes much larger than the first and hence $Z(\omega) \approx -j/\omega C$ and $I(t) \approx E(t)\ \omega C/(-j)$ and the phase shift is near 90°. This can be rationalized also by considering the fact that at low frequency the AC approaches a direct current and the impedance (the limit to current flow) is dominated by the capacitor, which does not permit direct current to flow.

At high frequency, the second component ($-j/\omega C$) becomes much smaller (goes to zero) than the first and hence $Z(\omega) \approx R$ and $I(t) \approx E(t)/R$ and the phase shift approaches 0°. In brief, at low frequency it is the capacitor, which impedes the current and at high frequency it is the resister that is the primary impediment to current flow. Again, if we consider that at high frequency, the capacitor does not charge fully and hence does not limit the flow of current, hence the resistor presents the limit to current.

An additional common electrical component is an inductor that describes the ability of a current passing through a coil (and creating a magnetic field) to induce a current in a second coil. This element is encountered infrequently in electrochemical experiments except, unintentionally, when a magnetic stirring plate is used close to the electrodes!

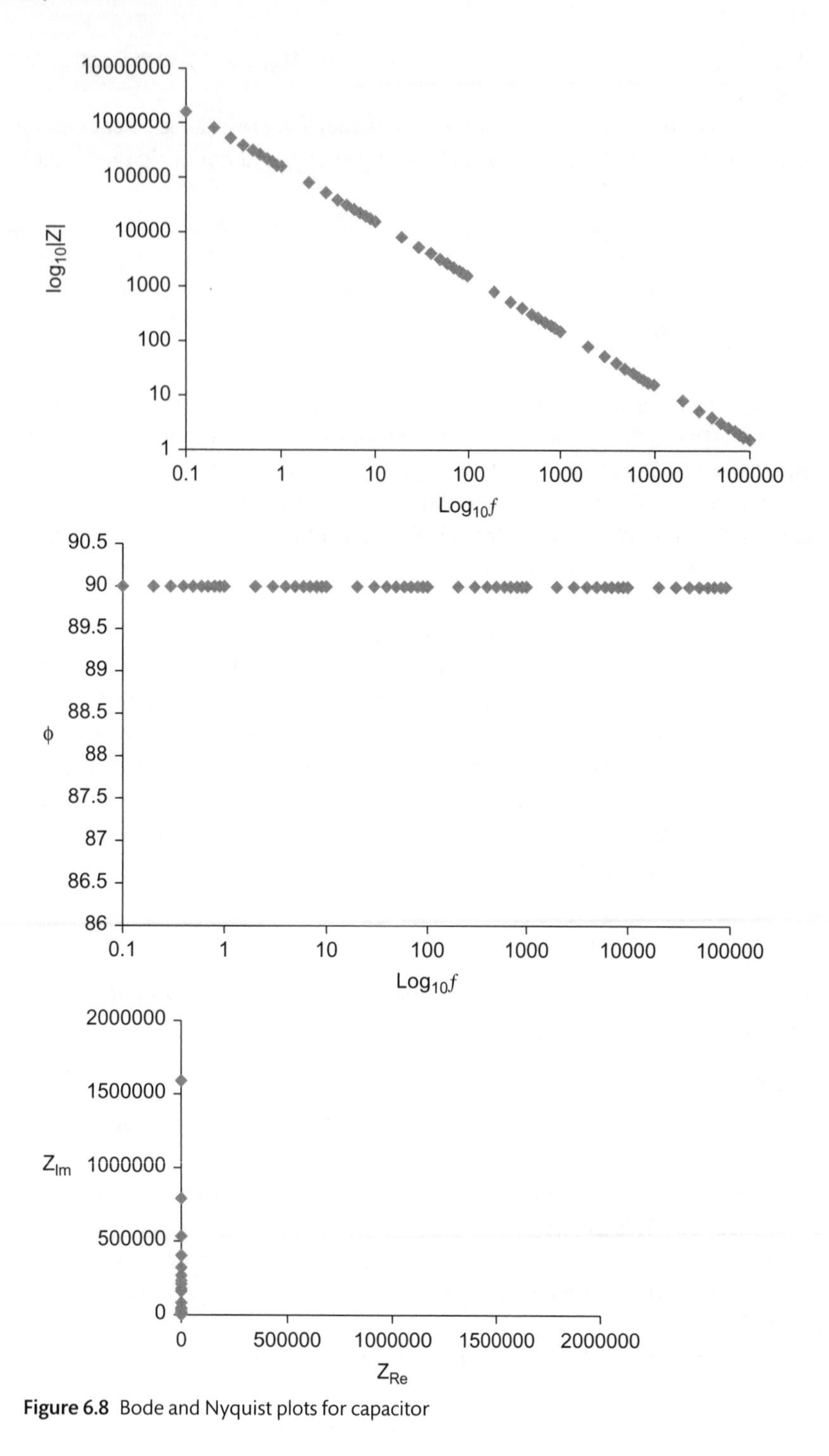

Figure 6.8 Bode and Nyquist plots for capacitor

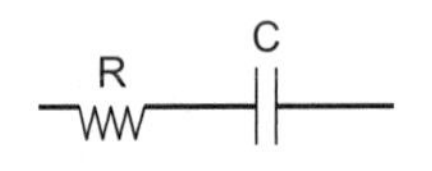

The situation at low frequency is opposite to that at high frequency and, hence at intermediate frequencies, when the impedance of the capacitor approaches that of the resistor, a situation intermediate between the extremes is found. Under these conditions, as the frequency is increased, the total impedance decreases concomitant with the phase shift resulting in a vertical line in the Nyquist plot with an intercept on the abscissa corresponding to the impedance of the resistor (Fig. 6.9).

The total impedance at any frequency can calculated by taking the absolute square (see Box 6.1).

$$|Z| = \sqrt{Z_{Re}^2 + Z_{Im}^2} = \sqrt{R^2 + \left(-\frac{j}{\omega C}\right)^2} = \sqrt{R^2 + \frac{1}{\omega^2 C^2}}$$

$$\tan\phi = \sqrt{\frac{Z_{Im}^2}{Z_{Re}^2}} = \sqrt{\frac{\left(-\frac{j}{\omega C}\right)^2}{R^2}} = \frac{1}{\omega RC}$$

6.5.7 **A parallel RC circuit**

We now consider a circuit in which a resistor and a capacitor are in parallel. In this case, the current can pass via the resistor or via the capacitor. At low frequency, that is, as we approach the DC limit, the capacitor is fully charged at all times and hence blocks current flow through it. Therefore, the current passes almost exclusively through the resistor and the overall impedance of the circuit is due to the resistor. In contrast, at high frequency, the capacitor does not charge fully at any time and offers essentially no impedance to current so therefore all current passes through the capacitor. The total current is the sum of the current passing through each component.

The impedance of a circuit in which a resistor and a capacitor are in parallel can be calculated using Kirchhoff's law also:

$$\left|\frac{1}{Z(\omega)}\right| = \sqrt{\left(\frac{1}{Z_1}\right)^2 + \left(\frac{1}{Z_2}\right)^2} = \sqrt{\frac{1}{R^2} + \frac{1}{\left[-\frac{j}{\omega C}\right]^2}} = \sqrt{\frac{1}{R^2} + \omega^2 C^2}$$

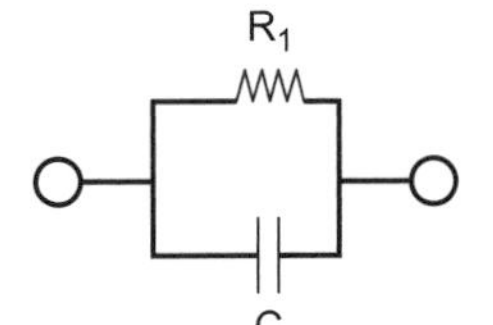

At low frequency, that is, ω is small, the relation simplifies to:

$$I(t) \approx E(t)\sqrt{\frac{1}{R^2} + \omega^2 C^2} = \frac{E(t)}{R}$$

Essentially all current passes through the resistor and the phase shift, ϕ, is 0°.

At high frequency, that is, ω is large, the relation simplifies to:

$$I(t) = E(t)\sqrt{\frac{1}{R^2} + \omega^2 C^2} \approx E(t)\sqrt{\omega^2 C^2} = \omega C\, E(t)$$

Essentially, all current passes via the capacitor and the phase shift is 90° (and varies at very high frequency as for a simple circuit containing only a capacitor: see Section 6.5.5).

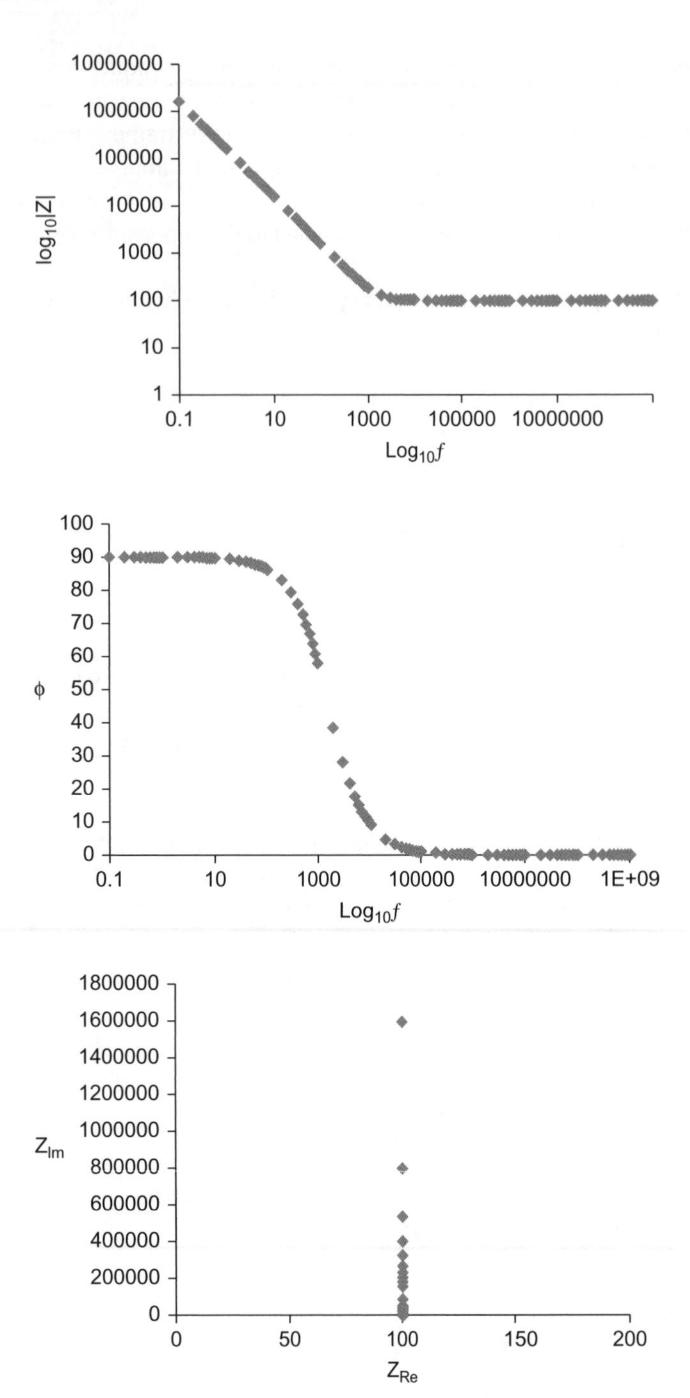

Figure 6.9 Bode and Nyquist plots for resistor and capacitor in series

At intermediate frequencies, the approximations cannot be made and the total impedance is calculated by:

$$\frac{1}{Z(\omega)}=\frac{1}{R}+\frac{\omega C}{-j}$$

$$\left|\frac{1}{Z(\omega)}\right|=\sqrt{\frac{1}{R^2}+\omega^2C^2}$$

and hence:

$$|Z(\omega)|=\frac{1}{\sqrt{\frac{1}{R^2}+\omega^2C^2}}$$

The phase shift is:

$$\tan\phi=-\omega CR$$

The Bode plots for a resistor and capacitor in parallel (Fig. 6.10) show the change in resistance and phase shift, ϕ. The Nyquist plot obtained from the polar coordinates ($Z(\omega)$ and ϕ) shows a semi-circle in which, at high frequencies, Z_{Re} is close to zero and at low frequencies the plot intercepts the abscissa at the impedance of the resistor. The frequency (ω) at which the maximum of the plot is reached $\left(\text{i.e.}\ \frac{dZ_{Im}}{dZ_{Re}}=0\right)$ is the inverse of the time constant of the circuit $\omega=1/RC$.

6.5.8 **EIS of a circuit in which a resistor is in series with a parallel RC circuit**

If we consider a resistor in series (R_2), with a capacitor (C) and a resistor (R_1) in parallel, then it is clear that the current must always pass through R_2. Therefore, the impedance is the sum of the impedance of R_2 and the combined impedance of R_1 and C.

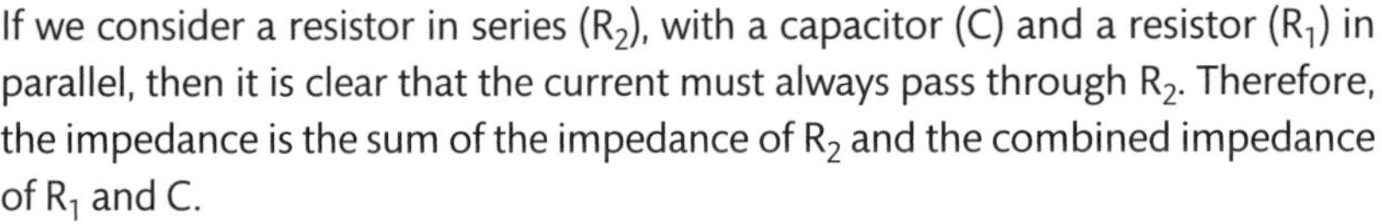

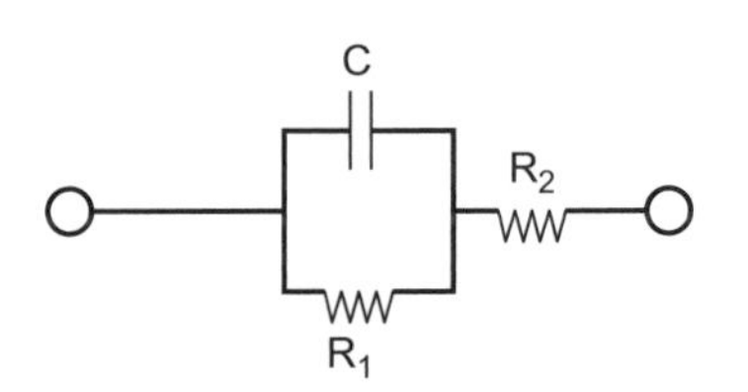

As discussed in Section 6.5.7, the contribution of the capacitor and resistor in parallel is:

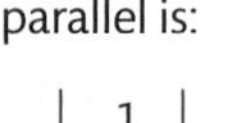

$$\left|\frac{1}{Z(\omega)}\right|=\sqrt{\left(\frac{1}{Z_1}\right)^2+\left(\frac{1}{Z_2}\right)^2}=\sqrt{\frac{1}{R_1^2}+\frac{1}{\left[-\frac{j}{\omega C}\right]^2}}=\sqrt{\frac{1}{R_1^2}+\omega^2C^2}$$

And hence total impedance of the circuit is:

$$|Z(\omega)|=\sqrt{R_2^2+\frac{1}{\frac{1}{R_1^2}+\omega^2C^2}}=\sqrt{R_2^2+\frac{1}{\frac{1+\omega^2C^2R_1^2}{R_1^2}}}=\sqrt{R_2^2+\frac{R_1^2}{1+\omega^2C^2R_1^2}}$$

At low frequencies the impedance of the capacitor approaches infinity and hence the current will pass primarily through R_1 and R_2 and the total impedance will be $Z=R_1+R_2$. This can be seen in the equation as the denominator of the second term approaches 1. At high frequencies the denominator of the second term approaches infinity and hence the whole of the second term approaches 0 and the impedance of the circuit is $Z=R_2$ since the current passes primarily through

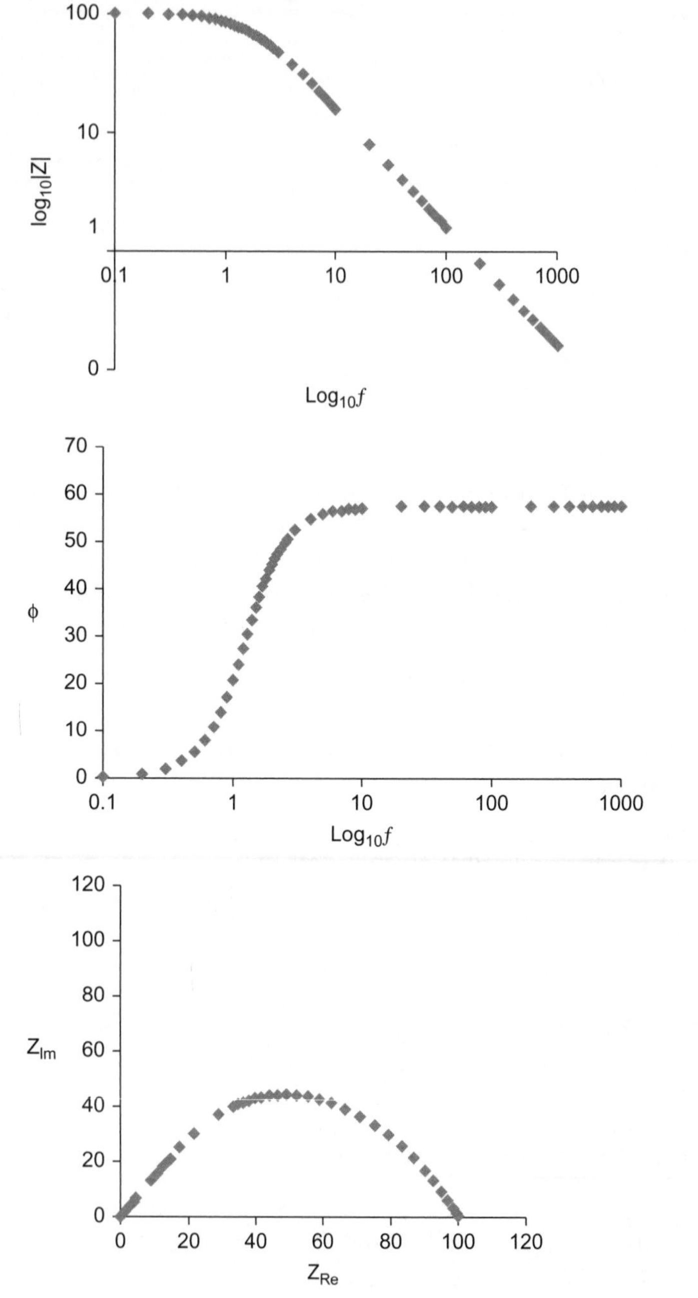

Figure 6.10 Bode plots and corresponding Nyquist plot for a resistor and capacitor in parallel

the capacitor and R_2. At intermediate frequencies, however, the impedance of both R_1 and C are similar and the current will pass through both elements.

The phase angle requires that the impedance be divided into its real and imaginary components:

$$Z(\omega) = R_2 + \frac{1}{\frac{1}{R_1} - \frac{j}{\omega C}} = R_2 + \frac{1}{\frac{\omega C - jR_1}{\omega C R_1}} = R_2 + \frac{\omega C R_1}{\omega C - jR_1}$$

The Bode plots for a resistor in series with a capacitor and a resistor in parallel show the change in resistance and phase shift, ϕ. The Nyquist plot obtained from the polar coordinates ($Z(\omega)$ and ϕ) shows a semi-ellipse in which at high frequencies Z_{Re} is at R_2, the series resistance. At low frequencies, the plot intercepts the abscissa at the impedance of the resistors in series ($R_1 + R_2$). The frequency (ω) at which the maximum of the plot is reached $\left(\text{i.e. } \frac{dZ_{Im}}{dZ_{Re}} = 0\right)$ is again the inverse of the time constant of the R_1C component of the circuit $\omega = 1/R_1C$.

We can increase the complexity of a circuit such as shown in Section 6.5.9; however, in each case, we simply calculate the total impedance and phase shift using Kirchhoff's laws and the experimentally determined Nyquist plot will show a series of parallel RC components each with their own time constants.

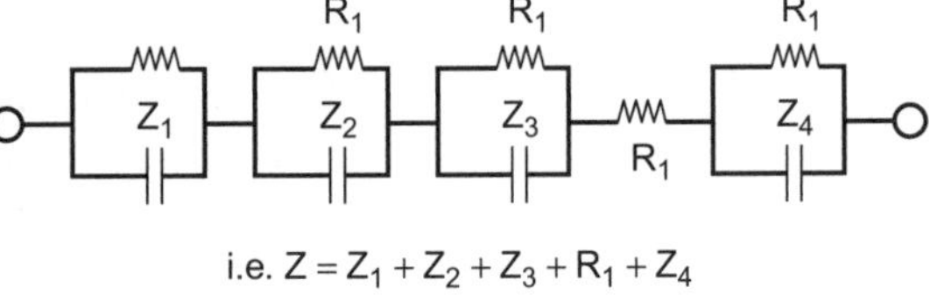

6.5.9 **Randles equivalent circuit**

It is easy to forget that the electrochemical cell, containing our compound of interest in electrolyte, is part of an electrical circuit. The double layer at the electrodes behaves like a capacitor, the electron transfer between a species in solution and the electrode gives rise to Faradaic (charge transfer, see Chapter 4 and the Tafel plot) resistance, and the electrolyte itself behaves like a resistor (cf. ***iR*$_u$** drop). Furthermore, **Warburg impedance** arises due to the limits on current imposed by mass transfer of redox active species to and from the electrode and, unfortunately, there is no equivalent solid state electrical component to mimic this kind of impedance. Each of these components can be modelled in the Randle's equivalent circuit as a combination of resistors and capacitors in series and in parallel; the electrolyte solution resistance (R_s) is in series with the double layer capacitance (C_{dl}), which is in turn in parallel with a charge transfer resistance (R_{CT}) and Warburg impedance (W) as shown in Fig. 6.11.

The total impedance in the Randles circuit is $|Z|^2 = Z^2_{Re} + Z^2_{Im}$

where $Z_{Re} = R_s + R_{ct}\left(1 + \frac{\lambda}{\sqrt{2\omega}}\right)$ and $Z_{im} = \frac{R_{ct}\lambda}{\sqrt{t\omega}}$

and $\lambda = \frac{k}{\sqrt{D}}$, k = chemical reaction rate and D is the diffusion coefficient:

The relative parameter for charge transfer more generally is $\lambda = \frac{k_f}{\sqrt{D_O}} + \frac{k_r}{\sqrt{D_R}}$

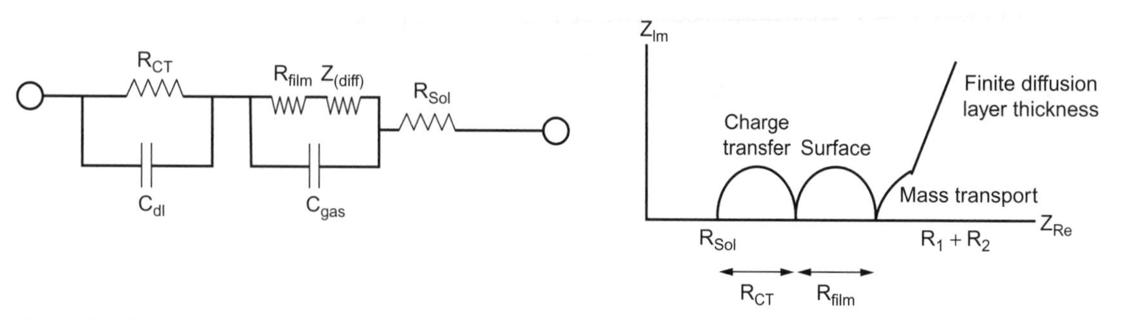

Figure 6.11 Randle's equivalent circuit and Nyquist plot for an electrochemical cell. Note that slope of the line showing mass transfer control is 45°.

A key challenge in modelling chemical systems is that the movement of redox active species by diffusion and the development of concentration gradients, for example, the kinetics of oxygen reduction in a solid state electrolyte, gives rise to an impedance component, Warburg impedance $(W) = Z(\omega) = \frac{R_{ct}}{\sqrt{t\omega}}\lambda$

The Nyquist plot obtained for a circuit involving an electrode in contact with an electrolyte containing a redox active species that is a gas in equilibrium with the solution is shown in Fig. 6.12.

Impedance spectroscopy finds broad application in the characterization of solid state devices but also in the study of, for example, polymer modified electrodes. For example, if the potential is varied close to the $E_{½}$ of the polymer film then ions will move in and out of the film to compensate for changes in charge. As the frequency increases the movement of ions is not fast enough to keep track with the change in redox state and the current response will vary (phase shift will increase).

6.5.10 **Final remarks**

The importance of batteries and fuel cells in modern life cannot be understated. In this chapter, only the briefest of overviews of this amazing field can be given, however, the general concepts discussed are applicable to almost all systems. In

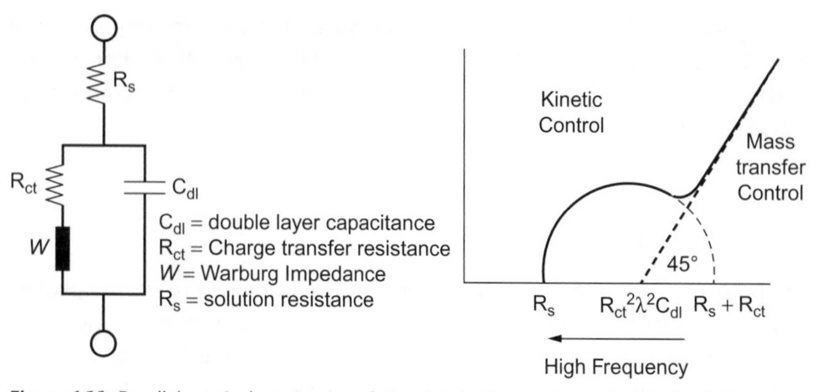

Figure 6.12 Equivalent circuit and corresponding Nyquist plot for O_2 reduction in a solid state electrolyte. Note that slope of the line showing finite diffusion is 45°.

contrast to chemical and biological solution-based applications of electrochemistry, where the interface is exposed and accessible to spectroscopic methods, in devices such as batteries, fuel cells, polymer solar cells, OLEDs, and so on, the interface is inaccessible and often changes in structure during use. This aspect, above all else, makes studying such systems highly challenging and EIS is an essential tool in the study of complex electrical devices. However, do not forget that EIS is extremely simple in a practical sense—we apply a sinusoidally alternating voltage with a specific frequency and measure the maximum current and the phase shift between current and potential that results from this. We repeat this experiment over a wide range of frequencies and from these two values we can generate plots such as the Nyquist plot.

The challenge comes in the interpretation of the data obtained and as a last word in this book, we should remind ourselves repeatedly that designing an equivalent circuit or a good kinetic model that fits our data well does not, in itself, mean that model is the correct one!

6.6 Summary

This chapter should provide you with an understanding of:

- the basic components of a battery and fuel cell
- the characteristics and balance of requirements in the design of batteries
- the concept of impedance due to a resistor and a capacitor
- prediction of the shape of Bode and Nyquist plots for a given circuit
- calculation of the impedance of a simple circuit
- the concept of an equivalent circuit
- the concept of Warburg impedance and its relation to concentration polarization

6.7 Exercises

6.1. Consider the disposable and rechargeable batteries in terms of the Nernst equation and electrodes of the 1st, 2nd, and 3rd kind. Explain why the two types of batteries function in different ways.

6.2. Consider the disposable battery based on nickel and gold (which although not likely to be a commercial success is conceivable).

a) Write down the battery in cell notation and suggest an appropriate electrolyte and electrodes.

b) Consider the choices of electrolytes, what difference would using KCl and K_2SO_4 make in terms of cell potential (calculations are not necessary in this answer but you should consider the redox potentials for the reactions)?

c) Would it be most efficient to make the electrodes equal in mass? Justify your answer.

6.3. Consider the nickel-cadmium battery: $Cd(s)|Cd(OH)_{2(s)}|KOH(aq)||NiO(OH)_{(s)}|Ni(OH)_{2(s)}|Ni(OH)_{2(s)}|Ni(s)$

a) What is the maximum power (in Watts) that this battery could provide with a 1000 Ω resistor?

b) If the maximum current were drawn, how long would the battery operate for (assuming the anode is 1 g and is the limiting factor)?

c) Will the battery EMF drop off suddenly or slowly at the end of its life?

6.4. Describe the procedure by which an impedance spectrum is recorded.

6.5. For the circuit shown, calculate the impedance (using Kirschoff's laws) at:

a) low,

b) high frequency, and at

c) intermediate frequencies.

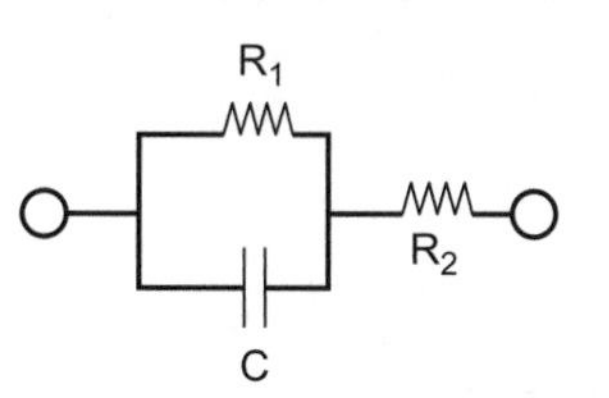

Glossary of terms

Admittance is the inverse of impedance, that is, the ability of a circuit component to allow current to flow.

Brownian motion is the random movement of particles in solution first noted by Robert Brown.

Carnot's theorem states that Carnot engine has the maximum efficiency possible when operating between to heat reservoirs. $W_{thermal}$ is the heat added to a system, T_c and T_h are the temperatures of the hot and cold reservoirs, respectively.

Le Chatelier's principle states that a system at equilibrium when disturbed will adjust its composition spontaneously to reach a new equilibrium position.

Electrode loosely refers to the wire, rod, or disc and so on used in an electrochemical experiment. The 'electrode' is better defined as the sum of the solid material and the solution and species it is in contact with.

Electrochemical area is not the same as geometric area and is a measure of the effect area of an electrode with respect to access to electroactive species.

Equilibrium potential is used synonymously with 'formal potential' or $E^{o}_{½}$, but differs due to differences in the diffusion coefficients of the oxidized and reduced forms of a redox active species.

Exchange current density, j_0 is the current per unit area passing across the solid liquid interface in each direction at the equilibrium potential.

Exergonic/Endergonic An exergonic reaction proceeds spontaneously as written while an endergonic reaction is not spontaneous; the terms are analogous and related to exothermic and endothermic, respectively, for changes in enthalpy during reactions.

Fermi level In metallic electrodes is the highest energy occupied electronic levels in an electrode.

Fugacity of a gas is the equivalent of the *activity* of species in solutions.

Impedance is labelled Z_{Re} and Z_{Im} as well as Z' and Z'', respectively.

'*iR* drop' or iR_u is the product of the resistance of the solution between the WE and RE and the current at the WE (and CE). This may seem counterintuitive but the current flows at the WE and hence even if it comes via the CE, the current is still flowing 'between' the WE and RE.

Laminar flow is the movement of solvent as thin sheets of solvent that glide over each other.

Net current density (*j*) has units of A cm^{-2} and is independent of electrode area.

OCP, 'open circuit potential' is the potential measured when the circuit contains a voltmeter only and current cannot pass via an external circuit.

Reaction coordinate diagrams represent how high in energy a system is as a reaction proceeds. The abscissa is typically, but not always, the bond length(s) that change the most in the reaction.

Solvation shell, the semi-ordered layer of solvent around a solute. The solute orients such that its dipole moment minimizes the charge density of the dissolved ion or molecule.

Tension An alternative term for voltage, especially when discussing power lines.

Transfer coefficient or symmetry factor, α, is typically 0.5 for reversible electrochemical systems. The symmetry factor is only applicable for single elementary step reactions, while the transfer coefficient is used for multi-step reactions.

Turbulent flow is the movement of solvent in a jumbled manner.

Warburg impedance Resistance to current due to limitations in mass transfer rates: redox active species have to diffuse to the electrode and the rate of diffusion is dependent on concentration gradients.

Appendix

A.1 Activity of ions

In non-electrolytes (i.e. only neutral molecules) the activity (a) is essentially the same as the molality ($a = m/m^{o}$) (where m^{o} is the standard molality (1 mol kg^{-1})).

In solution, the coulombic interaction between ions is strong and hence this approximation only holds when the ion concentration is low ($< 10^{-3}$ mol kg^{-1}).

So, what is activity? Activity is an expression of the difference between the actual chemical potential (μ) and the chemical potential under standard conditions (μ^{o}) where the ions behave ideally (i.e. no coulombic interactions).

$$\mu = \mu^{o} + RT\ ln\ a$$

where $a = \gamma x\ (m/m^{o})$, γ is the activity coefficient (and is not a fixed constant but varies with conditions between 1 (ideal) and 0 (infinite interaction)).

Therefore:

$$\mu = \mu^{o} + RT\ ln\ a = \mu^{o} + RT\ \ln\{\gamma x\ (m/m^{o})\} = \boldsymbol{\mu^{o} + RT\ ln\ (m/m^{o}) + RT\ ln\ \gamma}$$

'$\boldsymbol{\mu^{o} + RT\ ln\ (m/m^{o})}$' is the chemical potential of an ideal dilute solution.

A.2 Debye-Huckel Limiting law

The Dybe-Huckel limiting law allows the activity coefficient (γ) to be calculated at very low concentrations.

$$\log \gamma = -|z_{+}z_{-}|A(I/m^{o})$$

$A = 0.509$ in water at 25°C but depends on the relative permittivity of the solution and the temperature.

$I = ½\ \Sigma\ z_i^2 m_i$, that is, the ionic strength of the solution, which for a single salt → $I = ½\ (m_{+}z_{+}^{2} + m_{-}z_{-}^{2})$

A.3 Thermodynamics of electrochemistry

Standard free energy changes (ΔG) and enthalpy change (ΔH): The formation of ions in solution is associated with enthalpy (ΔH_f^{o}) and Gibbs free energy (ΔG_f^{o}) changes.

The equilibrium constant, K, is related to the free energy change by the equation:

$$RT\ ln\ K = -\Delta G^{o}$$

The standard enthalpy for a reaction such as:

$$Ag(s) + ½Cl_2(g) \rightarrow Ag^{+}(aq) + Cl^{-}(aq)$$

is the sum of the enthalpies of the formation of the individual ions.

$$\Delta H^{o} = \Delta H_f^{o}\ (Ag^{+}, aq) + \Delta H_f^{o}\ (Cl^{-}, aq) = -61.58\ kJ\ mol^{-1}$$

However, although the enthalpy change for the reaction overall has a physical meaning (i.e. it can be measured), the enthalpy change for each of the individual reactions has little real meaning:

$$Ag(s) \rightarrow Ag^{+}(aq) + e^{-} \quad Cl_2(g) + 2e^{-} \rightarrow 2Cl^{-}(aq)$$

The problem is that a point of reference is needed. This problem is overcome by defining the enthalpy change for the formation of H^+ (i.e. $\Delta H_f^o(H^+, aq)$) as being zero at all temperatures.

$$\Delta H_f^o(H^+, aq) = 0$$

Therefore, for:

$$\tfrac{1}{2}H_2(g) + \tfrac{1}{2}Cl_2(g) \rightarrow H^+(aq) + Cl^-(aq) \quad \Delta H^o = -167.16 \text{ kJ mol}^{-1}$$

$$\Delta H^o = \Delta H_f^o(H^+, aq) + \Delta H_f^o(Cl^-, aq) = \Delta H_f^o(Cl^-, aq)$$

Hence:

$$\begin{aligned}\Delta H_f^o(Ag^+, aq) &= \Delta H^o - \Delta H_f^o(Cl^-, aq) = -167.16 \text{ kJ mol}^{-1} - (-61.58 \text{ kJ mol}^{-1}) \\ &= -105.58 \text{ kJ mol}^{-1}\end{aligned}$$

The same approach can be used for the Gibbs free energy change for the formation of ions. The entropy of the H^+ ions in water is also taken to be zero.

Further reading and online resources

The goal in preparing this book was to provide a first step up to students, both under- and post-graduate level, to enter the amazing field of electrochemistry. The length of such a book inevitably means that we cannot journey too deeply into any particular points. However, there are several books that I have found to be especially useful in going deeper, and I recommend that any researcher that will spend a significant amount of time on the topic should have on their desk.

Sawyer, D. T., Sobkowiak, A., Roberts, J. L., Jr., 1995, *Electrochemistry for Chemists, 2nd ed.*, Wiley.

Bond, A., 2002, *Broadening Electrochemical Horizons, Principles and Illustration of Voltammetric and Related Techniques*, Oxford: Oxford University Press.

Bard, A. J., and Faulkner, L. R., 2001, *Electrochemical Methods: Fundamentals and Applications, 2nd ed.*, New York: Wiley Global Education.

Four general text books that I recommend both to the novice and as a refresher course in physical and analytical chemistry are:

Harris, D. C., 2016, *Quantitative Chemical Analysis, 2015, 9th ed.*, Freeman.

Skoog, D. A., West, D. M., Holler, F. J., and Crouch, S. R., 2013, *Fundamentals of Analytical Chemistry, 9th ed.*, Brooks/Cole.

Atkins, P. W., and de Paula, J., 2009, *Atkins' Physical Chemistry, 9th ed.*, Oxford: Oxford University Press.

Anslyn, E. V., and Doherty, D. A., 2006, *Modern Physical Organic Chemistry*, Sausalito, California: University Science Books. (Chapter 7 for Transition state and reaction theory.)

Several primers that are recommended for further reading beyond this book are:

Maczek, A., and Meijer, A., 2017, *Statistical Thermodynamics*, Oxford: Oxford University Press, for a solid introduction to Maxwell's relations.

Compton, R. G., and Sanders, G. H. W., 1996, *Electrode Potentials*, Oxford: Oxford University Press.

Fisher, A. C., 1996, *Electrode Dynamics*, Oxford: Oxford University Press.

Brett, C., and Brett, A. M. O., 1998, *Electroanalysis*, Oxford: Oxford University Press.

There is a series of papers in a specific issue of the *Journal of Chemical Education* in addition to others that are usually available through university (online) libraries. Although some are now quite old, they provide an excellent starting point for going deeper into the subject and are well recommended including those dealing with basic concepts,[1,2] thermodynamics,[3] the Nernst equation,[4] cyclic,[5–7] and pulse voltammetry,[8] electrochemical mechanisms,[9,10] anodic stripping voltammetry[11], chronocoulometry,[12] the Tafel plot[13–15], electrocatalysis,[13–16] modified electrodes,[17] electropolymerization and redox polymers,[10,15,18] electrosynthesis,[15,16] fuel cells,[19] impedance spectroscopy,[20] the glucose biosensor,[21] and organic redox reagents.[22] There are also some articles that are of general interest one of which is a discussion[23–25] of sign conventions (something I have happily ignored in this book). One would do very well to start building a general reference library in electrochemistry with these works and although there is quite a lot of overlap, one also gets the chance to read from different perspectives.

1. Faulkner, L. R., 1983, Understanding electrochemistry: some distinctive concepts. *J Chem Educ.* 60: 262.
2. Frumkin, A., Petry, O., and Damaskin, B., 1970, The notion of the electrode charge and the Lippmann equation. *J. Electroanal. Chem.* 27: 81.
3. Battino, R., Wood, S. E., and Strong, L. E., 1997, *A Brief History of Thermodynamics Notation*. 74: 304.
4. Vidal-Iglesias, F. J., Solla-Gullón, J., Rodes, A. et al., 2012, Understanding the Nernst equation and other electrochemical concepts: an easy

experimental approach for students. *J. Chem. Educ.* 89: 936.

5. Elgrishi. N., Rountree, K. J., McCarthy, B. D. et al., 2018, A practical beginner's guide to cyclic voltammetry. *J. Chem. Educ.* 95: 197.
6. Van Benschoten, J. J., Lewis, J. Y., Heineman, W. R. et al., 1983, Cyclic voltammetry experiment. *J. Chem. Educ.* 60: 772.
7. Mabbott, G. A., 1983, An introduction to cyclic voltammetry. *J. Chem. Educ.* 60: 697.
8. Osteryoung. J., 1983, Pulse voltammetry. *J. Chem. Educ.* 60: 296.
9. Dryhurst, G., Nguyen, N. T., Wrona, M. Z. et al., 1983, Elucidation of the biological redox chemistry of purines using electrochemical techniques. *J. Chem. Educ.* 60: 315.
10. Genies, E. M., and Tsintavis, C., 1985, Redox mechanism and electrochemical behaviour of polyaniline deposits. *J. Electroanal. Chem.* 195: 109–28.
11. Wang, J., 1983, Anodic stripping voltammetry: an instrumental analysis experiment. *J. Chem. Educ.* 60: 1074.
12. Anson, F. C., Osteryoung R. A., 1983, Chronocoulometry: A convenient, rapid and reliable technique for detection and determination of adsorbed reactants. *J. Chem. Educ.* 60: 293.
13. Fang, Y. H., and Liu Z. P., 2014, Tafel kinetics of electrocatalytic reactions: from experiment to first-principles. *ACS Catal.* 4: 4364.
14. Costentin, C., Drouet, S., Robert, M., et al., 2012, Turnover numbers, turnover frequencies, and overpotential in molecular catalysis of electrochemical reactions. cyclic voltammetry and preparative-scale electrolysis. *J. Am. Chem. Soc.* 134: 11235.
15. Hendel, S. J. and Young, E. R., 2016, Introduction to electrochemistry and the use of electrochemistry to synthesize and evaluate catalysts for water oxidation and reduction. *J. Chem. Educ.* 93: 1951–6.
16. Francke, R. and Little, R. D., 2014, Redox catalysis in organic electrosynthesis: basic principles and recent developments. *Chem. Soc. Rev.* 43: 2492.
17. Bard, A. J., 1983, Chemical modification of electrodes. *J. Chem. Educ.* 60: 302.
18. Genies, E. M., and Syed, A. A., 1984, Polypyrrole and poly-*n*-methylpyrrole, an electrochemical study in an aqueous medium. *Synth. Metals* 10: 21.
19. Smith, D. E., 1983, Thermodynamic and kinetic properties of the electrochemical fuel cell. *J. Chem. Educ.* 60: 299.
20. Scipioni, R., Jørgensen, P. S., Graves, C. et al., 2017, A physically-based equivalent circuit model for the impedance of a $LiFePO_4$/graphite cylindrical cell. *J. Electrochem. Soc.* 164: A2017–30.
21. Wang, J., 2001, Glucose biosensors: 40 years of advances and challenges. *Electroanalysis* 1312: 983.
22. Connelly, N. G., and Geiger, W. E., 1996, Chemical redox agents for organometallic chemistry. *Chem Rev.* 96: 877.
23. Woolf, A. A., 1990, Electrochemical conventions: responses to a provocative opinion (3). *J. Chem. Educ.* 67: 26010.
24. West, A. C., 1986, Electrochemical cell conventions in general chemistry. *J. Chem. Educ.* 63: 609.
25. Anson, F., 1959, Electrode sign conventions. *J. Chem. Educ.* 36: 394.

Index